# Mobile VPN: Delivering Advanced Services in Next Generation Wireless Systems

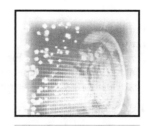

# Mobile VPN: Delivering Advanced Services in Next Generation Wireless Systems

## Alex Shneyderman and Alessio Casati

WILEY

Wiley Publishing, Inc.

Publisher: Robert Ipsen
Editor: Carol Long
Developmental Editor: Emilie Herman
Managing Editor: Micheline Frederick
Text Design & Composition: Wiley Composition Services

This book is printed on acid-free paper. ∞

Published by Wiley Publishing, Inc., Indianapolis, Indiana

Published simultaneously in Canada

Limit of Liability/Disclaimer of Warranty: While the publisher and author have used their best efforts in preparing this book, they make no representations or warranties with respect to the accuracy or completeness of the contents of this book and specifically disclaim any implied warranties of merchantability or fitness for a particular purpose. No warranty may be created or extended by sales representatives or written sales materials. The advice and strategies contained herein may not be suitable for your situation. You should consult with a professional where appropriate. Neither the publisher nor author shall be liable for any loss of profit or any other commercial damages, including but not limited to special, incidental, consequential, or other damages.

For general information on our other products and services please contact our Customer Care Department within the United States at (800) 762-2974, outside the United States at (317) 572-3993 or fax (317) 572-4002.

**Trademarks:** Wiley, the Wiley Pubishing logo and related trade dress are trademarks or registered trademarks of Wiley Publishing, Inc., in the United States and other countries, and may not be used without written permission.

Wiley also publishes its books in a variety of electronic formats. Some content that appears in print may not be available in electronic books.

*Library of Congress Cataloging-in-Publication Data:*

ISBN 0-471-21901-0

Printed in the United States of America

10  9  8  7  6  5  4  3  2  1

# Dedication

Alex Shneyderman:
To my wife Olga, who has truly become my (better) half, and to our first ten years together; to my son Mark, who will not need to know about the complexity of the high-tech world around him (unless he chooses daddy's path); and to my parents for being there for me.

Alessio Casati:
To my wife Lina Maria, who helped me throughout this effort with her patience, support, love, and understanding. To my family and loved ones.

# Acknowledgments

As virtually any undertaking of such magnitude, this book would not be possible without the participation and generous help of a number of people and organizations. First and foremost, we would like to thank Igor Faynberg who introduced us to Wiley Publishing and guided us through the writing and material selecting process, providing invaluable technical and organizational insights along the way.

We would like to thank our colleagues at Lucent Technologies for their support and help with the details of both business and technological aspects of MVPN and advanced IP services. Especially, we would like to thank Pete Mccann, Dawn Mckenna, Mikko Kiukkanen, and Peter Hsu. Special thanks goes to Mike Grinn, Said Tatesh, and Benny Siman-Tov for taking their time to proofread many of the book chapters and provide us with valuable comments.

This book would not be possible without the help and participation from our editing team at Wiley, which always demonstrated the highest level of professionalism while providing technical reviews and suggestions on how to improve the quality of our text and for guiding us through the publishing process. We would like to especially thank Emilie Herman and Felicia Robinson for their contributions.

Finally, we would like to thank our families—especially Alex's wife Olga, son Mark, and parents Boris and Vieta, and Alessio's wife Lina and parents Natale and Luisella for their constant support, warm encouragement, and patience during our temporary exile from public life, frequent mood swings, and periods of being mentally withdrawn, which unfortunately often accompany undertakings of such magnitude.

**Alessio Casati**
**Wootton Bassett, UK**
**Alex Shneyderman**
**Wayne, New Jersey**

# Contents

# Foreword

Mobility and security are becoming dominant themes in the Internet of the new millennium. These factors present many challenges and many opportunities for healthy growth of the Internet. Mobility heightens the need for security, just as it brings into sharp focus the need for applying other technologies like service discovery, location management, tunneling, and remote data management. Taken together, these technologies represent a fundamental restructuring of previous approaches for continued growth of the Internet.

Tunneling alone is powerful enough to effect such a restructuring, but network designers rarely create solutions that involve pure tunneling. Instead, the tunnel management is equipped with many and sundry techniques for deciding when to enable, start, and stop the use of the tunnel. All of these techniques amount to an expression of some policies, which are motivated by the problem that the tunneling solution is intended to solve. Two very prominent tunneling policy regimes are Mobile IP and VPNs. The former is intended to create useful routes so that data can be delivered to whatever foreign domain the mobile node should visit. The latter is supposed to create useful routes so that data can be delivered securely to whatever foreign domain at which a portable device may become situated at.

When portable devices become wireless, they (and their users!) become much more mobile. Soon, traditional methods for managing VPNs will be viewed as inadequate. Mobility techniques, perhaps those derived from Mobile IP, or Mobile IP itself, will be adapted to fulfill the needs of devices whose communications are protected by VPNs. This will create the Mobile VPNs described in this book.

I believe this book will help the practitioner understand the many related wireless, mobility management, and security technologies that can be used to create MVPNs. Undoubtedly, there will be many ideas for solutions, depending on the background and goals of the designers. And yet, security is very tricky, so care and study are crucial for success.

Nothing beats experience, and it is important to learn from the experiences of others, understanding why their solutions succeeded or failed. Moreover, vendors for existing products that are related to MVPNs will surely try to take advantage of opportunities for extending their product lifetimes. Thus, we are likely to see many different variations upon the technologies and themes presented in this book.

My own work with Mobile IP has given me the opportunity to meet and work with Alessio Casati and Alex Shneyderman. Alessio and I made an initial effort to make more tunneling tools available to Mobile IP, namely GTP, the tunneling mechanism of choice within GPRS domains. That was after I had grown to enjoy Alessio's forthright assessments of various developments on the IETF mobile-ip mailing list. That same spirit of forthright assessment and technical presentation will make this book accessible and valuable for the many readers interested in solutions providing mobility and security. I wish them, and this book, the best of luck to find that readership.

**Charles E. Perkins**
**Nokia Fellow, Nokia Research Center**
**Mountain View, CA**

# Preface

*"Alice came to a fork in the road. 'Which road do I take?' she asked. 'Where do you want to go?' responded the Cheshire cat. 'I don't know,' Alice answered. 'Then,' said the cat, 'it doesn't matter.'"*

***Lewis Carroll,** Alice in Wonderland*

In June 2002 the worldwide number of wireless subscribers finally reached a long anticipated 1 billion, beating even the most optimistic analyst's estimates by almost a year. The growth rate of Internet hosts is not far behind, pushing 200 million at the time of this book's publication. The important conclusions, however, can be drawn from the percentage of total wireless users who are also using the Internet. According to the latest analyst's research, close to 80 percent of Internet users are also cellular wireless subscribers, and the percentage of business users for both is about 40 percent.

Not surprisingly, there is a growing interest in wireless data for both commercial and residential applications—especially in the context of the next generation systems such as CDMA2000, GPRS, UMTS, EDGE, and WLAN. One of the most prominent recent trends in commercial landline data communications is the use of virtual private networking (VPN) to gain secure connection to remote private networks over public shared infrastructures such as Internet. The obvious next step is to apply the benefits of landline VPN technologies to mobile environments, creating Mobile Virtual Private Networks (MVPNs) that allow secure pervasive private communications over a variety of shared mobile networks offered by wireless operators and wireless Internet service providers (WISPs).

## What This Book Is About

We are at the stage in the wireless data communications life cycle when most of its underlying technologies have reached maturity or at least have stabilized in standards. Third-generation wireless systems, which defined new services and provided high-speed data and multimedia support, have been standardized and are being rapidly deployed all over the world (albeit with some expected early-growth problems, unfortunately, perversely associated to financial issues due to unprecedented financial speculation and overexcitement around new technologies at the end of the past millennium). All of these systems include support for packet-switched data communications, as opposed to their predecessors, which relied on circuit-switched technologies. The important aspects of these systems include not only the potentially higher throughput, which is often relatively modest during initial rollouts and under certain circumstances, but also availability, always-on capabilities, higher efficiency, and a strong foundation for the delivery of new services. Packet data (covered in Chapter 4), effectively brought wireless networks one step closer—in functionality, throughput, and resource utilization—to wireline data networks while preserving and enhancing one of its most important features of wireless communications: *mobility.*

Additionally, advancements in virtual private networking technologies allowed the easy use of public shared wireline infrastructures such as the Internet to securely transmit private data traffic, thus extending the reach of private networks by allowing remote users to connect to remote resources, information, and services. These advancements have mainly occurred in the past 3 years, so we believe now is the best time to come up with a comprehensive guide on the subject that would allow the reader to study it on any desired level of complexity and perhaps help in addressing research and development efforts in the most appropriate direction.

The focus of the book is remote access to private networks in the wireless environment achieved by the use of technology called *Mobile Virtual Private Networking.* Most of the discussion is centered on the second- and third-generation cellular systems—supporting packet-switched data communications—and Wireless LANs. These systems, including CDMA2000, GPRS, and UMTS, are being deployed at this writing and are expected to proliferate throughout the world during the next few years. With this text we strived to provide a comprehensive guide on the subject complete with background on both wireline data and cellular wireless communications. The book touches on special topics such as support for Mobile VPN over integrated cellular/WLAN infrastructures and the support of advanced Mobile IP services. In the second part of the book we introduced a new approach to classification of VPN and other private network access methods used in GPRS, UMTS, and CDMA2000 and discussed a variety of other complex issues in the field of wireless data. We supplemented the material with a healthy amount of practical network design scenarios and real-life examples (specifically in Chapters 6, 7, and 9). The book also provides a look at the open issues and future trends in wireless data communications in general and at private network access over untethered medium in particular.

## How This Book Is Organized

The book is structured in two main sections. Part One provides necessary background material to prepare you for the discussion on MVPN, such as an overview of related standard bodies, a practical guide to standard documentations, and a discussion of the current state of standardization in the area of wireless data. The second part focuses on the subject at hand—Mobile VPN—and other topics pertinent to providing advanced packet data services in mobile environments.

Chapter 1 introduces the mobile VPN business case and the MVPN market segment, and it gives you an overview of the wireless standardization process and a brief "standard document retrieval manual" template. Here we also provide the explanation of standards organizations' hierarchy, timelines, and milestones, as well as the organizations' involvement with particular technologies and systems. Chapter 2 provides a tutorial on relevant topics in data networking and communications, such as MPLS and IP security, while Chapters 3 and 4 discuss radio interface fundamentals and details of various cellular systems of interest, as well as background information on both circuit- and packet-switched wireless data provided within second- and third-generation systems frameworks.

Part Two provides an in-depth discussion of VPN support in the mobile environment, and it contains useful information about how wireless data and specific IP services are supported in current and next-generation cellular wireless systems, such as various flavors of CDMA2000, GPRS, and UMTS. In Chapter 5, we transition from wireless to Mobile VPN (MVPN), while providing a discussion on the VPN taxonomies, underlying technologies, tunneling and security issues, and other VPN building blocks. Chapter 6 focuses on the properties of MVPN provided in GSM and UMTS cellular systems, while Chapter 7 covers MVPN services in the CDMA2000 family of systems. Chapter 8 briefly analyzes main types of the equipment involved in Mobile VPNs. Chapter 9 concludes the book by looking at future trends and forward-looking topics such as cellular/WLAN integration issues, Mobile VPN in converged networking environments, and services provided by Mobile Virtual Network Operators (MVNOs).

This book also includes appendixes with specific information about Mobile IP extensions and RADIUS attributes and a Bibliography that lists books and standard documents containing additional material on a variety of related subjects. We believe the resulting structure should provide a sufficient foundation to follow the discussion in the second half of the book, even for the readers with limited background in wireless and data communications.

## Who Should Read This Book

While this book provides an ample tutorial of wireline data and wireless communications, we did not want to write yet another book on data communications or the basics of cellular wireless systems. *Mobile VPN: Delivering Advanced Services in Next-Generation Wireless Systems* covers only relevant topics and disciplines comprising Mobile VPN and other advanced wireless data service. Consequently, this book requires a reader to have basic education in computer science and electrical engineering. We also assume you have some exposure to the basics of data communications and cellular wireless concepts, as well as an understanding of OSI model and TCP/IP, and link layer technologies such as PPP, ATM, and Frame Relay.

Reading this book does not require a previous understanding of IP tunneling, wireless data communications, Wireless LAN fundamentals (although knowledge of regular LAN technology is still required), VPN concepts, data networking security, or next-generation cellular systems and services taxonomies. Those readers with a good understanding of these and other related technologies will find in-depth discussions on these and other related subjects especially useful. Sometimes even familiar and well-studied and documented subjects are presented in a new light of application to wireless and mobile VPN.

We tried to make this book useful for a wide audience of professionals and students in wireline and wireless data communications and networking, as well as representatives from other professions, such as investors and financial analysts wishing to gain proficiency in this subject. In particular: Chapters 6, 7, discussing the implementation of Mobile VPNs within particular wireless systems, and Chapter 9, discussing the future directions of wireless data services, offer comprehensive case studies and real-life examples of virtual private network deployments and address practical network design issues such as IP addressing strategies, VPN optimization methods, and practical wireless IP security. This makes this book especially useful to both wireless and wireline data network architects, IT managers, network administrators, and even corporate security officers involved in implementing of real-life systems.

Engineers and product managers involved into implementation and definition of products supporting MVPN functionalities will benefit from our in-depth descriptions of the functionalities of various elements required in next-generation cellular and WLAN systems to offer MVPN services. The analysis of business issues and future trends of MVPN field provided in the first and last chapters should be of special interest to product managers and other professional involved in the product feature decision making and market forecasting.

## From Here

We believe that wireless data services and Mobile VPN in particular have a great potential to not only improve business productivity and give services providers new revenue opportunities, but also offers significant improvements in our everyday lives. The successful deployment of com-plex third-generation wireless data systems and services like Mobile VPNs will entirely rest on the shoulders of thousand of professionals in the fields of wireless and data communications, as well as the next generation of engineers and managers completing their studies. With their help, new and exciting products will be rolled out by equipment manufacturers and new services will be offered by operators. We hope this book will help this group in both understanding of the technology and its application and practical deployment to bring the progress in wireless communication to a new level and continue the remarkable run it has enjoyed for the last two decades.

After all, unlike Alice and the Cheshire cat, we believe that it does matter which way you choose. There are many forks on the road of wireless communications and we hope that this book will not only help the readers to decide where they want to go, but also guide them through many difficult turns.

PART

# One

# Wireless Data Fundamentals

# Introduction to MVPN

Contracting margins and revenues per user, cost-based competition, and focus on customer retention rather than acquisitions are all signs that mobile telephony is not likely to show significant revenue growth—comparable to that enjoyed over the past decade—over the next several years. Service providers are therefore forced to look for innovative ways to invest in new technologies, which can potentially become the next growth enablers. For instance, in recent years much attention has been paid to "mobile Internet" services, which are believed to pose a significant revenue-generation potential for service providers.

This belief was in part responsible for the massive investment in spectrum for next-generation radio access technologies, with the potential to support higher data rates for mobile Internet services, commonly known as *third generation* (3G). More recently, service providers have recognized that Internet access per se may not justify the significant investments they made. As a result, the search is back on for innovative ways to generate revenues by using the new service capabilities offered by the deployment of packet-data-based systems such as General Packet Radio Service (GPRS), Universal Mobile Telecommunication System (UMTS), or CDMA2000 (CDMA stands for Code Division Multiple Access). So far, it is apparent that the most promising kind of services mix traditional mobile voice capabilities and new location-based and messaging services. Such systems must provide users personalized and predictable access to private networks where they can belong to communities of interest for both business and leisure, such as corporate networks or instant messaging groups.

The value of such networks to the customers appears to be strictly related to:

- Ensuring secure network access with predictable performance
- Making sure that access to such networks is exclusive to members with appropriate permissions.

These service requirements are compelling service providers to use Mobile Virtual Private Networking (MVPN), which we define as the emulation of private secure mobile data networks over generally insecure shared mobile and wireless facilities. This definition is based on a number of assumptions:

- Data user mobility is defined as uninterrupted connectivity or the ability to stay connected and communicate to a possibly remote data network while changing the network access medium or points of attachment.
- Despite MVPN service is usually provided over wireless media, and in fact, this book is written about VPN implementation over various wireless access systems. We make clear distinction between "mobile" and "wireless," since these terms have different meanings and we believe that for our purposes "mobile" is more accurate and inclusive (see Chapter 5 for more discussion on wireless versus mobile).
- The term "wireless facilities" refers to current and future generations of cellular systems of interest such as Global System for Mobile Communications (GSM), CDMA2000, Time Division Multiple Access (TDMA), and UMTS, wireless networks such as Wireless LANs (WLANs), and overlay wireless packet data systems such as GPRS.

A simple visualization of MVPN can be found in Figure 1.1, which shows secure tunnels connecting a mobile device with a variety of private networks over multiple shared public networks, such as the Internet and an arbitrary cellular wireless system or WLAN network.

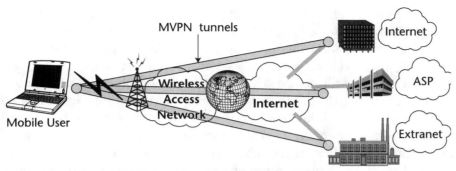

**Figure 1.1**    Example of Mobile Virtual Private Networking.

In this chapter we concentrate on business and standardization issues as an introduction to topics addressed in the rest of the book. The first half of this chapter discusses the MVPN business case and marketplace, explaining what benefits this technology can bring to service providers and their customers. We start with the discussion of pervasive mobility and its consequences, moving on to MVPN history and business case. The section ends with an overview of MVPN market segments and stakeholders. The second half of the chapter examines the current wireless data standardization status and trends and provides the reader with a reference to the standard documents usage and retrieval from various standard body repositories. The reader will become familiar with standards organizations such as 3GPP, 3GPP2, and the Internet Engineering Task Force (IETF), along with their standardization processes. An understanding of what a given standard body does within the landscape of mobile networking will be helpful to the reader for the remainder of the book.

# The Era of Pervasive Mobility

We are fortunate to be witnessing the beginning of an era of *pervasive mobility*, when access to information resources will not be determined by the availability or type of network access technology but rather by factors such as the desire, necessity, and eligibility to obtain information or services. Information and services will be requested and accessed not only by individuals but also by virtual and physical entities such as automated manufacturing processes, "smart" vending machines, information-collecting devices such as utility meters, intelligent cash registers, highway toll stations, security systems, and medical equipment. (See Chapter 9 for some anticipated next-generation services scenarios.) Remote network access service characteristics will not be dependent on geographical location, but rather on the existence of proper roaming and service agreements between home and visited data networks, allowing for home service profile retrieval into foreign networks. When proper agreements are in place, mobile entities or individuals will be able to receive services identical to those available in their "home" networking environments while roaming foreign networks.

## Pervasive Mobility Drivers

So what drives the need for pervasive mobility, or permanently available uninterrupted on-demand connectivity? The most important drivers are productivity gains via advancing IT technology, the rise of the Internet, the ever-increasing speed of evolution of mobile devices, cellular and noncellular network coverage, and plummeting costs of cellular wireless service.

### Increase in Productivity

The changing role of information technology in corporations and institutions throughout the world was responsible for major productivity gains in the workplace during the last decade of the twentieth century. That was, of course, accompanied by the rise of the Internet, which brought together masses of information and united disparate communities of interest all over the world. However, massive computerization also brought total dependence on computing and information resources often available only at a limited number of select locations, such as corporate headquarters or data centers. The newly available services, so indispensable to users in their offices, are more and more often requested from remote locations such as satellite or home offices, customer sites, and from the road. These needs in turn drive demand for global network roaming and ubiquitous remote network access. The uncertainty of the location where the user will require access imposes the need for mobile (dynamic) private connectivity to the home network to be available throughout a wide area (also referred to as ubiquitous access).

### Mobile Device Evolution

It is hard to underestimate the role of personal communication and computing devices—such as Personal Digital Assistants (PDAs), smart mobile phones, and laptop computers leaving the factory with multiple built-in wireless interfaces—in the evolution of mobile communications. Plummeting prices, increasing user-friendliness, and feature richness are now making these devices not only increasingly available and desirable for ever-increasing groups of mobile professionals, but often indispensable.

The latest example of such devices is a slew of PDAs and PDA-based mobile phones that are approaching earlier-generation desktop computers in memory and processing power. Manufacturers like HP, Toshiba, Sony, Nokia, and Palm are leading this trend. These small, low-powered wonders can support multiple modes of wireless communications (Infrared, Bluetooth, WLAN, and GPRS or CDMA2000), VPN clients, and a micro-browser bundled with the operating system. This combination of features makes them ideal for secure access to remote corporate networks.

### Cellular Systems Advances

The third driver for pervasive mobility is the rapid build-out and consolidation of cellular systems resulting in more and more uniform wide-area wireless coverage and increasingly inexpensive services. Cellular coverage has become so widespread and inexpensive that it is driving other technologies out of business. This resulted in a situation where *alternative* wide-area wireless access systems, such as satellite, had been deemed unnecessary by a paying public whose low demand forced them into bankruptcy or niche markets. Good examples of those are now defunct satellite operator Iridium and Inmarsat, which is forced to specialize on maritime navigation. In fact, cellular service in some areas, ranging from highly saturated European and Japanese markets to developing nations completely lacking landline infrastructure, became so ubiquitous and affordable that it began to replace landline phone service.

## Mobile Lifestyles and Workplaces

These and other technological advances account for the rise of pervasive mobility, which in turn continues to bring profound ongoing changes to our society, lifestyle, and workplace in how we communicate, how we receive information and news, and how we process the information. The 90's and the first years of the new millennium were essential in the formation of mobile technology. The way our society now uses cellular wireless networking and, specifically, wireless data technologies such as short messaging service (SMS) is the best indicator of these changes. From teenagers using SMS for "secret" communications to professionals in Japan using i-mode devices for banking services, wireless data users are fast approaching voice users in numbers and generated revenues. In fact, it is expected that wireless data will become the leading driving force of telecommunications in the coming decade.

Stopping short of producing yet another set of arguments in favor of this technology to add to the numerous articles and books already written, let's now turn our attention to the main subject of this book. In our view the next "hot" mobile technology will involve wireless data as a foundation for specialized services like location-based service, private environments, mobile-commerce (m-commerce), and MVPN. We discuss these services further in Chapter 9. For now, we start our discussion of MVPN by taking a brief tour of VPN history, then outlining Mobile VPN business case.

## Background on VPN

Virtual Private Networks were originally defined and first applied in voice communications. For years, phone companies delivered voice services using what they called "Virtual Private Networks," despite the fact that there wasn't—and still isn't—much that is "virtual" about them. In fact, even today just about any software-defined user group provisioned over any physical medium is considered VPN by the phone companies. The term is still in use even though Public Switched Telephone Network (PSTN) facilities are owned by the phone companies, thus making the technology essentially a private network used for offering user group services.

With the rise of data communication, the term VPN was adopted by data networking industry and was given a new, more accurate meaning. So-called traditional data VPNs were initially created with dedicated link layer networking technologies such as Frame Relay PVC (Permanent Virtual Circuit) or ATM VC (ATM stands for Asynchronous Transfer Mode) links, established between individual hosts or networks. In roughly 10 years following the advent of these technologies, data VPNs typically have been implemented in this fashion with the main goal of replacing less efficient private networks based on dedicated end-to-end leased facilities.

**NOTE** Interestingly, later on both ATM and Frame Relay were gradually reclassified as private networking technologies, mainly on the grounds that while these networks were shared, they nevertheless were privately owned. This also made sense for marketing purposes; ATM and FR services could be equated to those available through the use of truly private, dedicated technologies, such as leased lines, thereby promoting these new data transport methods.

As the use of the public Internet Protocol (IP) networks such as the Internet quickly gained public interest and market acceptance, a new generation of VPN services based on network layer technologies has been introduced to the market. Like traditional VPNs, IP VPNs utilize shared facilities to emulate private networks and deliver reliable, secure services to end users. During the initial IP VPN technology trials, equipment manufacturers and standards organizations such as the IETF came up with a number of encapsulation and encryption techniques (more on those in Chapter 2) in an effort to deliver on the promised cost advantages and complexity reduction [Yuan2001], without compromising security requirements many potential VPN customers have. The proprietary mechanisms like Layer Two Forwarding (L2F) devised by Cisco and Point-to-Point Tunneling Protocol (PPTP) introduced by Microsoft include such early examples. Ultimately, the industry settled on the use of standard based technologies such as IPSec, L2TP, Generic Routing Encapsulation (GRE), and Multi-Protocol Label Switching (MPLS), among others (details are also in Chapter 2). Common authentication and accounting methods largely based on the RADIUS protocol previously defined to satisfy the demand for centralized subscriber management in the remote dial-up industry were also selected and standardized for use with IP VPN. Mobile wireless VPNs are the latest members of this group.

# MVPN Business Case

Mobile VPN is a data service that can be provided within any system or network supporting authenticated (most often wireless) access to a data network. Let's look at MVPN business case as a combination of VPN and wireless data business cases and analyze its value for operators, in the form of revenue and marketing potentials, and customers, as a vehicle for delivery of new services and functionality. Based on our findings, we will look at MVPN market from both service provider and customer perspectives and evaluate the MVPN benefits and values proposition for specific customer and provider segments.

## Moving to Mobile VPN

The Internet, now accessible from almost anyplace where telephone lines, cellular service, or satellite services are available, has fundamentally changed the way we communicate and access information and services. The Internet is rapidly becoming the medium of choice for business communications. However, it is a public shared network, whereas business communications requires private secure facilities. That means if the Internet is to be used for private communications, the user information transported over it must be somehow secured.

Suites of networking and security technologies were devised in response to this requirement and quickly became popular methods for conducting private communications over the Internet or any other shared IP networking medium. These were known as IP VPN technologies. In the course of wireless networks' development, the requirement of mobility became more and more stringent in the provisioning of IP VPN services. This fostered research, standards efforts, and the development of MVPN technologies in the industry. Today, operators are preparing business plans and architectures to support a variety of MVPN offerings to serve the needs of their business and institutional customers.

## Wireless Communications with MVPN

For wireless operators deploying latest-generation cellular systems based on packet-switched data such as GPRS and CDMA2000, and especially those targeting *business* customers for significant portion of their revenue stream, the importance of services based on MVPN technologies is hard to underestimate. For operators, MVPN is not only one of the required technologies for business customers' private network access but also a foundation for other services requiring interaction with private networks such as m-commerce, virtual presence and gaming applications, and multimedia applications (which includes Voice over IP-based services).

The benefits of deploying Mobile VPNs for businesses and institutions include:

- Uninterrupted, media and location-independent connectivity to private networks
- Seamless private network access mobility
- Connectivity to a particular Internet service provider (ISP) or application service provider (ASP)

- Mobile remote access outsourcing possibilities
- Secure m-commerce enabler
- Constant remote-workers reachability
- Higher cost-effectiveness

As a result, businesses, which already had a positive experience with wireline VPN services, are now looking to wireless operators for extending these services into wireless environments. In our view, during the next few years as the latest generations of cellular systems and other wireless technologies take off, an enormous market opportunity awaits wireless carriers who can meet demands for services requiring private network access.

## MVPN as a Differentiation Tool

Mobile VPN is a powerful differentiation tool especially for service providers dedicating a significant portion of their offerings to serving business customers. But why is differentiation so important? During the last few decades, we witnessed the cellular communications gradually rise from luxury item and status symbol to necessary tool and then abruptly become a commodity. Commoditization is undesirable for any industry. But it is also a natural part of almost any product life cycle and is no stranger to the wireless telecommunications. As Internet-based Web services boomed in recent years, ubiquitous and fast Internet access became an immediate goal of the wireless data industry as well. However, wireless data services consisting mainly of Internet access are considered as generic as today's cellular voice services, and face price-based competition that is already hurting wireless voice revenues. Subscribers will not have enough incentive to stay with a particular carrier for reasons other than pricing. The situation is further complicated by the ease of switching from one carrier to another, known by operators as *customer churn*, which has plagued the wireless industry since its inception.

MVPN seems to be one of the likely answers to these problems. Since MVPN technology is highly customizable and can be implemented in different flavors to accommodate different customers needs, the MVPN service offerings can also be packaged and marketed differently by different wireless operators. That means they would be able to offer more "sticky" or unique services, and therefore prevent price-driven customer churn. This is especially true for large institutions and enterprises expected to make unique MVPN services from different wireless operators an integral part of their IT infrastructures.

# Mobile VPN Market and Stakeholders

Having touched on the MVPN overall value proposition and its significance in the wireless market, let's now turn to MVPN market segments. The
MVPN market, like any other market, consists of buyers (private network
access customers) and sellers (wireless carriers and other service providers)
who engage in transactions concerning a particular product or product
category (private network access in mobile environment in our case)
[Kotler1999]. Let's now look at each group, taking into account the benefits
of MVPN and different deployment strategies more suitable to different
customers and providers.

## MVPN Service Providers

MVPN service providers can be loosely classified into three major groups:

- Wireless operators
- Mobile Virtual Private Operators (MVNOs) and other subcontractors
  and resellers
- Wireline Internet service providers (surprise!)

The wireless operators group includes carriers, offering both actual
network-based MVPN services as well as the business-quality Internet
access with properties suitable for stable end-to-end VPN to be created
between the customer's mobile equipment and the customer's VPN gateways. (See Chapter 5 for details on the difference between end-to-end and
network-based VPN types.) Wireless carriers are by far the largest MVPN
service provider group. This is not surprising, since the wireless carriers
own both the spectrum licenses and the radio infrastructure. For network-
based MVPN offerings, wireless carriers would have to establish proper
agreements with the enterprises and institutions defining trust relationships, legal liabilities, quality of services, availability, and other parameters. If the enterprise chooses an end-to-end VPN method, the role of
wireless carriers would be to support an end-to-end MVPN-compatible IP
addressing scheme based on the use of publicly routable IP addresses
or appropriately designed private address translation mechanisms. Of
course, in the case of end-to-end VPN, wireless carriers might be bypassed
altogether and might not even be aware of private communication taking
place over their infrastructure, since by the nature of end-to-end VPN,
packets transmitted between tunnel endpoints are encrypted and bear only
end-to-end significance. This makes network-based service more attractive
to wireless carriers; they can introduce a multitude of offerings with high

revenue-generating potentials and institute more control over the mobile subscribers.

The second category of potential MVPN providers includes Mobile Virtual Network Operators (MVNOs) and other wireless service subcontractors, such as those engaged in infrastructure-sharing agreements (the radio access network and the spectrum license normally belong to their business partners). This group should be especially interested in network-based VPN services for the same reasons as regular wireless carriers. In addition, when the end-to-end VPN option is selected by the MVNO's corporate customers, MVNO's role would be much restricted and revenues would be likely marginal, since the middle-man role the MVNO targets would be quite limited. So MVNOs are even more likely to strive to add as much value as possible by implementing intelligent services in the network. (More discussion on MVNOs is provided in Chapter 9.)

The last service provider group, which might be considered unlikely by some, includes traditional wireline Internet service providers. This group can participate in providing network-based MVPN service through agreements with wireless operators, which often allow the latter to use highly developed IP infrastructures of the former. The benefits of offering MVPN service for this group mostly lies not in new revenue-generating capabilities but in product line extension—that is, in augmenting their traditional wireline offerings with newly available MVPN options. This allows wireline ISPs to become one-stop service providers for their traditional customers regardless of the network access method (wireless or wireline). We also need to stress that *wireline* service providers in some countries are starting to drive the deployment of a WLAN-based hot-spot coverage infrastructure, thus seeking as much independence as possible from *cellular* wireless operators and at the same time trying to compete with them in wireless high-speed data services.

## MVPN Customers

The benefits of deploying Mobile VPNs are as significant for customers as they are for service providers. MVPNs provide remote workers with constant, media-independent connectivity to corporate networks or to the ISPs and ASPs of their choice. MVPNs enable corporations to outsource mobile remote access, and in some cases can completely replace *wireline* remote access infrastructures—thereby eliminating the costs of purchasing and supporting the remote access equipment while allowing private networks to maintain full control over user address assignments, authentication, and security (see Chapter 5).

Let's take a closer look at the potential MVPN users and their requirements. Generally, they can be classified into the following main categories:

- Small businesses
- Large enterprises
- Institutions, both government and academic
- Applications service providers (ASPs)

The following sections look at each user category in more depth.

### Small Businesses

The main motivation for small businesses to use MVPN is primarily convenience and its cost-cutting abilities. MVPN is generally used by small businesses for remote access to centralize information resources, email access, and the monitoring of certain events, such as medical monitoring and utility billing. MVPN service for small business is more likely to be achieved via end-to-end connectivity, which does not require the establishment of complex agreements with wireless carriers. Instead, the responsible personnel must make sure that employees and partners are provided with the business class wireless Internet access with proper qualities to support end-to-end MVPN service.

Another reason why end-to-end VPN is more likely to be utilized within this segment is its relative ease of implementation and low price. To support this service, the remote workers must be provided with mobile devices equipped with off-the-shelf or proprietary VPN clients and security software and equipment such as IPSec protocol stacks and RSA SecureID cards. Often clients are bundled with operating systems—for example, IPSec clients are bundled with Microsoft Windows 2000 used with laptop computers (more in Chapter 8).

### Enterprises

The main reason larger enterprises would be interested in MVPN is the potential productivity gains and increased personnel reachability. Cost cutting and ease of deployment will remain secondary issues. In an enterprise, MVPN services are most likely outsourced via an agreement or a number of agreements with wireless carriers, which are responsible for providing remote employees and partners of an enterprise with specific types and classes of MVPN services. In this situation, all types of MVPN can be used with equally good results as long as they satisfy cost, security, convenience, ease of support, and other requirements of an enterprise.

Generally, large enterprises are not as cost-sensitive as small businesses or government institutions. They often desire state-of-the-art services for their remote mobile workers—such as high-speed mobile data access and special security arrangements—which require a variety of MVPN technologies; network-based ones often being the most suitable. Usually, enterprise IT departments require to be involved in many aspects of the services provided by carriers, which, for instance, would allow them to retain control over policy provisioning, authentication, or IP address assignment. In these situations, open management interfaces, as well as carefully structured provisioning arrangements, are critical.

### *Institutions*

Government and other public institutions might be interested in MVPN services for reasons different from those motivating the private sector. For example, telecommuting is encouraged by the US government primarily not for cost-cutting reasons but to reduce pollution by eliminating daily travel to work. Home offices are becoming more and more popular with many public institutions. This trend, however, requires large-scale remote access mechanisms such as landline IP VPN combined with MVPN for workers on the road.

Service requirements of a government institution often can be rather unpredictable and unexpected (anyone in the private sector who has dealt with government customers can attest to this), often for a good reason or at least with good intentions. For that reason, flexibility in MVPN offerings and technologies should be the key when dealing with public institutions of various sizes and functions. For example, security requirements can often be very strong and far exceed those customary for private sector.

On the other hand, government institutions are often required to be especially cost-conscious and must structure their spending according to yearly plans.[1] This prompts the use of very detailed service level agreements between government institutions and wireless carriers defining all the up-front prices and the services these prices would buy. Offering compulsory VPN service also relieves an institution from the responsibility of participating in VPN setup, provisioning, and maintenance—all of which can be outsourced to wireless carriers and their partners.

Academic and medical institutions are usually bound by similar goals of careful use of often substantial resources and the desire to use the latest technology available to achieve certain unique objectives such as support for telecommunications research projects or remote patient diagnostics. MVPN requirements for this group often have the attributes of both large

---

[1] 5-year plans were used in some countries in the past without much success.

enterprises and public institutions. For this reason, the approach that should be taken by wireless carriers should be one of diversity. Often the service presenting the right features might consist of a combination of offerings and unique arrangements, such as a combination of end-to-end and network-based VPN services, use of granular per-flow policies, and unique arrangements for traffic differentiation and service bundling.

### Application Service Providers

This class of MVPN customers will grow as the wireless carriers take advantage of application packages offered by their wireline partners or content providers. These players must rely on dedicated private virtual networks so that the control of access to the services they offer can be easily enforced, and business class and predictable network access makes the user experience in accessing the services they offer uniform. ASP VPN offerings also come with advanced accounting features, so that the wireless provider and the partners can mutually exchange correlated traffic and content usage data and apply to these discounting policies and offer services like trend analysis and customer-behavior monitoring. These MVPNs allow members to access ASP services and offer subscription-based access to a host of services in a service bundle without forcing customers to perform individual authentication procedures.

## Wireless Data Standards

It is no doubt that interoperability and multivendor-based solutions is one of the key market requirements in telecommunications industry today. In particular, compliance to standards has always been one of the main requirements to MVPN solutions, since they potentially span multiple networks (access, transit ISP, customer) and are inherently bound to interoperability between wireless access devices and the access network infrastructure to allow for global roaming. It is therefore necessary to have a good understanding of the standard bodies we will be referencing in the book.

The need to produce standards for the advent of the next generation of wireless systems prompted the foundation of a number of Standard Definition Organizations (SDOs) during the last few years. Third-generation wireless systems requirements were originally defined by the ITU (International Telecommunications Union, a United Nations-associated organization) within the IMT-2000 framework (International Mobile Telecommunications). Aside from defining some technological and spectrum requirements for the radio transmission technologies that could be considered candidate for 3G services, the IMT-2000 framework defined service requirements such as the support of global roaming.

This forced all the parties (manufacturers and operators) involved in standards setting to evolve the standardization bodies at a global level. The result was the creation of the Third-Generation Partnership Project (3GPP) organization and later the foundation of a mirror organization (not without a touch of irony in the name) called 3GPP2. Before we define the scope and organization of these SDOs, we should step back and look at the landscape of cellular-industry-related standards organizations before the advent of the 3GPP and 3GPP2.

## Regional Standards Organizations

Aside from international organizations such as the ITU, which had almost no influence on the definition of cellular wireless systems, each region in the world had its own standard-setting bodies devoted to this technology. The GSM system in the 900- and 1800-MHz spectrum was defined by European Telecommunications Standards Institute (ETSI) and later was also adopted by the Telecommunications Industry Association (TIA) T1-P1 committee in North America for the 1900-MHz spectrum (dedicated to PCS, or Personal Communications Services). TIA has also defined a host of other cellular systems in the North American region, both analog and digital. Japan, in contrast to the rest of the world, defined its own digital cellular system called Personal Digital Communications (PDC). The Japanese standard bodies are the Association of Radio Industries and Businesses (ARIB) and the Telecommunication Technology Committee (TTC). These standards bodies also influence the decision making in the rest of the Pacific Rim—with the exception of Korea and China, which have their own organizations (the Korean Telecommunications Technology Association and China Wireless Telecommunication Standards Group, respectively).

Each of these organizations were defining regional standards incompatible with standards defined by other organizations, with the exception of the GSM 1900-MHz system, that would allow GSM 900 and 1800 customers to roam to North America using a tri-band phone. It was clear that this model could not work anymore, as the need to standardize 3G systems guaranteeing global roaming arose. In an ideal world, a single new organization defining a single system for the whole world would have been a logical solution. Of course, we don't live in a perfect world. Instead, the spectrum allocation for the 3G services in Europe and Japan was in the 1900- to 2100-MHz region, which was already partially used by the PCS services in the North American region. This situation, together with a different migration paths to 3G and different core network technologies used in the existing European GSM and the North American CDMA systems, led to even more profound mutual incompatibilities, not entirely by chance, right from the birth of two distinct SDOs for third-generation systems: 3GPP and 3GPP2.

## 3GPP

3GPP is an agreement between regional telecommunications standards bodies known as *Organizational Partners*. Currently, the 3GPP Organizational Partners are ARIB, China Wireless Telecommunication Standard Group (CWTS), ETSI, T1, Telecommunications Technology Association (TTA), and Telecommunication Technology Committee (TTC). 3GPP was created in December 1998 when the partners signed the Third-Generation Partnership Project Agreement. A company can be a member of 3GPP providing that it meets the rules defined for 3GPP membership.

A second category of partnership was created within the project: *market representation partners*. These are organizations and industry focus groups driven by objectives based on long-term needs of their member companies. At a certain stage, these partners decide it is important to have 3GPP hear their opinion as a group, rather than disseminate this opinion via the individual member companies. One of these groups, the 3G.IP industry focus group, has been particularly influential in driving the standardization of the evolution of the 3GPP system's specifications toward an IP-based, multimedia-capable system.

The intended purpose of 3GPP was to define technical specifications and technical reports for a 3G Mobile System based on an evolution of the GSM core network. This included the radio access technologies that were selected for 3G services based on the GSM core network: W-CDMA-based Universal Terrestrial Radio Access (UTRA) in its Frequency Division Duplex (FDD) and Time Division Duplex (TDD) modes. Later it became evident that it would make sense to extend the scope of 3GPP to include the maintenance and evolution of the Global System for Mobile communication (GSM) technical specifications and technical reports, and the related radio access technologies and services, General Packet Radio Service (GPRS) and Enhanced Data rates for GSM Evolution (EDGE). Now 3GPP has taken over the roles for which ETSI had been responsible.

The work of 3GPP is organized into Technical Standardization Groups (TSGs), which in turn are organized into Working Groups (WGs). The rules of operation of a TSG are specified in the technical report 3G TS 21.900 "Technical Specification Group Working Methods."

Following is the list of current TSGs groups comprising 3GPP:

- TSG-SA (System and Architecture) defines the systems aspects and coordinates the technical work of all other groups from a systems perspective. It includes five WGs:
  - SA1 handles requirements.
  - SA2 Systems handles architecture.
  - SA3 handles security.
  - SA4 Voice handles multimedia coding.
  - SA5 handles charging.

- TSG-CN (Core Network) specifies the evolution of the core network. There are five Working Groups in TSG CN:
  - CN1 addresses the protocols between the user equipment (UE, also known as terminal, mobile phone, or mobile station) and the core network [specifically the node in the core network that dialogues with terminals in order to manage UE mobility and allow the mobile station (MS) to set up and receive calls].
  - CN2 specifies the interaction of the mobile network with intelligent network functionality and services.
  - CN3 defines the interworking of the mobile network with external networks, such as the PSTN, or packet data networks, such as the Internet.
  - CN4 specifies core network protocols.
  - CN5 specifies application programming interfaces and protocols used to access network services from third-party application providers.

- TSG-RAN (Radio Access Network) defines the UMTS Terrestrial Radio Access Network (UTRAN). It is composed of four WGs:
  - RAN-1 is devoted to radio physical layer protocol specification.
  - RAN-2 handles the specification of the radio link layer.
  - RAN-3 defines the Iu interface (that is, the interface between the radio access network and the core network).
  - RAN 4 addresses pure radio aspects.

- TSG-GERAN (GSM Evolution RAN) defines specifications for the evolution of the GSM Radio Access Network. It is composed of five Working Groups:

- GERAN1 is devoted to radio aspects.
- GERAN2 addresses protocol aspects.
- GERAN3 is devoted to GSM Base Station Subsystem testing OA&M.
- GERAN4 specifies radio aspects of terminal testing.
- GERAN5 WG5 addresses protocol aspects of terminal testing.
- TSG-(T) (Terminals) specifies terminal aspects. It includes three WGs:
  - T1 addresses test specifications for interoperability
  - T2 specifies terminal capabilities
  - T3 specifies the UMTS Subscriber Identity Module (SIM), which is a chipcard that enables subscriber identity authentication, terminal portability, and execution of simple applications.
- A Project Coordination Group (PCG) has the role of determining rules of operation of the body and defining its working procedures.

3GPP specifications are delivered in *releases*. Initially, ETSI released specifications every year and assigned names accordingly. The first release of UMTS specifications (which was also a GSM specifications release, because of the role of GSM maintenance and evolution that 3GPP took over) was named Release 99. Later, as soon as the following release 2000 development plan had to be articulated, the decision was made to lift the constraint binding 3GPP specifications releases to a year and instead use functionality-based releases. 3GPP releases are now named with a release number different from the year the release was issued, starting from year 2000. The first release issued under this new naming convention was named Release 4, the second Release 5, and so on. The counter started from 4 because the specification version number was 3.$x$.$y$ (where $x$ and $y$ are generic-figure placeholders) for Release 99, and the decision was made to increment the first number in the version number of a specification at every release.

Release 99 defines the basic UMTS features associated with the circuit-switched and packet-switched services UMTS provides. Release 4 enhances the circuit services part of the system to use the latest developments in media gateways and media gateway controllers' technologies, and Release 5 introduces the support of multimedia services over the packet-switched part of the system.

Release 6 will introduce, among other features, multicast and broadcast capabilities that make the delivery of multicast content economically viable. Note that by pure coincidence the issue of a release happens every year, so the completion of the R5 specification is 2002, and Release 6 specification is expected to be completed in 2003.

## 3GPP Documents and Standardization Process

3GPP produces specifications that result in two sets in permanent documents: technical reports (TRs) and technical specifications (TSs). A TR is a permanent document that records a Working Group activity, such as the investigation on the feasibility of introduction of some feature in the specifications. A TS is the actual document specifying the behavior of network nodes and the definition of protocols used in 3GPP-compliant systems.

The three kinds of technical specifications are as follows:

- Stage 1 specifications outline service and functional requirements and are based on the input from the operators. SA1 is the WG within TSG-SA that normally generates all Stage 1 documents for 3GPP.

- Stage 2 specifications address system-level and architectural-level requirements that the protocols specified by 3GPP should meet. These are the documents where all the strategic directions and political decisions are formalized. Normally, SA2 within TSG takes the role of generating the most high-level Stage 2 documents, while when more specific competence is required on the protocol level, other WGs define Stage 2 documents.

- Stage 3 documents are the actual 3GPP protocols specifications.

Generating a TS is a formal process. In the first phase, interested companies, led by a document rapporteur or group of rapporteurs, contribute heavily to generate a first draft of the document. The document is generated by a WG by consensus. When the draft document is ready, a WG submits it to the TSG plenary for approval. After the approval, the document is promoted to a higher level of stability. The WG submits the document again to the TSG plenary, suggesting that it is stable enough to enter the change control phase. In this stage of a document's lifetime, companies can change the document only by submitting a formal change request (CR). A specification belongs at any given time to a release. When a release is "frozen," changes to the document can be approved only by general consensus or because there are serious system operation problems if the document does not change. A document can evolve over a number of releases, until 3GPP decides to withdraw a specification starting from a 3GPP release.

Documents (better known as temporary documents, as opposed to TSs and TRs, which are permanent documents) submitted by interested companies are discussed at WG meetings. A set of output documents is the result of consensus of the WG meetings. This set of documents is forwarded to the TSG plenary, where normally they are approved (or companies that feel their voice was not adequately heard at the WG level can ask for changes or the rejection of one or more of them). When documents are approved by a TSG plenary, the content is normally transferred to a permanent document. Thus, after every TSG plenary the version number of the documents under the control of a TSG changes.

## 3GPP2

3GPP2 is a collaboration agreement among standard definition organizations interested in developing specifications for the 3G systems evolving from an ANSI-41-based core network, set up in February 1999 for the same reasons that led to the creation of 3GPP to define specifications for 3G systems evolving from the map-based GSM core network. The Organizational Partners that are currently members of 3GPP2 are ARIB, CWTS, TIA, TTA, TTC, and Market Representation Partners (MRP).

Much like 3GPP, 3GPP2 felt the need to allow the market to bring organized input to their standardization activities. 3GPP2 Market Representation Partners (MRP) are organizations that can offer market advice to 3GPP2 and bring into 3GPP2 a consensus view of market requirements (e.g., services, features, and functionality) falling within the 3GPP2 scope.

Following is a list of current MRPs:

- CDMA Development Group (CDG)
- Mobile Wireless Internet Forum (MWIF)
- IPv6 Forum

In particular, MWIF has been proposing strongly the evolution of the 3GPP2 systems specification to an IP-based multimedia-capable system (much like 3G.IP in the 3GPP arena).

The operation of 3GPP2 is guided by the Steering Committee.

The actual work in 3GPP2 is performed by TSGs. The TSGs responsible for generating the technical specification documents of 3GPP2 are as follows:

- TSG-A (Access Network Interface) is responsible for the specifications of interfaces between the radio access network and core network, as well as within the access network for capabilities like intervendor handoff.

- TSG-C (CDMA2000) is responsible for the radio access part, including its internal structure, of systems based on 3GPP2 specifications. Specifically, it is responsible for requirements, functions, and interfaces for the CDMA2000 infrastructure and user terminal equipment. This includes radio layer 1, 2, and 3 specifications, mobile and base station performance and test specifications, support for enhanced privacy, authentication and encryption, and digital speech and video codecs. It also addresses the mobile station-to-adapter interfaces and other ancillary interfaces.

- TSG-N (ANSI-41, Wireless Intelligent Network) is responsible for the specifications of the core network part of systems based on 3GPP2 specifications. These include core network internal interfaces for call-associated and noncall-associated signaling, evolution of the core network for intersystem operation within the ANSI-41 family member, Virtual Home Environment (VHE) procedures, user identity module (UIM) support (detachable and integrated), and support for enhanced privacy, authentication, encryption, and other security aspects.

- TSG-P (Wireless Packet Data Networking) is responsible for the specifications of packet data networking for 3GPP2 systems. These include Wireless IP services (including IP Mobility Management), Wireless IP network architecture design, Voice over IP, public Internet and secure private network access, packet data accounting, multimedia, and quality of service (QoS) methods. This group is strongly influenced by MRPs like MWIF. An ad hoc *TSG All IP* has been created to satisfy the MWIF requirements for an All IP-based system.

- TSG-R (Interface of 3GPP Radio Access Technology to 3G Core Network evolved from ANSI-41) is responsible for the Inter-Working Function specification of Interface of 3GPP Radio Access Technology (i.e., UTRAN) to 3G Core Network to an evolved ANSI-41 core network. It also addresses handoff between cdmaOne and UTRA radio technologies and roaming between ANSI-41 and GSM core networks. TSG-S (Systems and Services Aspects) is responsible for the development of service capability requirements for systems based on 3GPP2 specifications. It is also responsible for architectural issues as required to coordinate service development across the various TSGs.

### 3GPP2 Documents and Standardization Process

3GPP2 produces technical specifications and reports similarly to 3GPP. TSG-S defines feature and system requirements. These, much in the same way as in 3GPP, are referred to as Stage 1 requirements. Technical specifications and reports are developed in the TSGs. The specifications are developed in two stages:

- Stage 2 is a high-level overview of the implementation of a feature or service in the 3GPP2 architecture, including message flow diagrams.

- Stage 3 is the text and the associated information for the final technical specification.

Once a specification or report is technically stable and complete, the TSG approves the document as *baseline text*. The document undergoes a verification and validation (V&V) process. Once this V&V process has been passed, the document can be approved for publication by the TSG.

After a TSG approves a document, it forwards the document to the 3GPP2 Secretariat. The 3GPP2 Secretariat opens a 15-day comment period. If no comments are received, the document is published as an official 3GPP2 publication. The Organizational Partners (TIA, TTC) can subsequently handle the document according to regional standards approval processes. Once this review is complete, any comments are sent to the originating 3GPP2 TSG. The document then undergoes an update process. The updated document is then base-lined, subjected to V&V, and approved by the TSG as necessary. The process described previously is then repeated.

## Internet Engineering Task Force

Since most data applications in wireless networks are IP-based, it comes as no surprise that the IETF and the protocols it specifies are becoming increasingly relevant to the wireless data industry. The IETF is organized in areas that organize technically related Working Groups. The current IETF areas are as follows:

- *Applications Area* deals with applications and application protocols like presence and instant messaging, network time protocol, calendaring, and scheduling.

- *General Area* addresses topics related to the general operation of the IETF, such as rules setting.

- *Routing Area* specifies routing protocols and their applicability.

- *Internet Area* defines IP protocol-related matters, such as the definition of its evolution, the support of network services such as PPP, and IP host configuration. Recently, it took on the role of specifying the Mobile IP protocol from the Routing Area, since Mobile IP is now perceived as a mobile remote IP network access technology, rather than a routing protocol.

- *Operations and Management Area* defines network management aspects and protocols, such as the well-known Simple Network Management Protocol (SNMP) and its evolution.

- *Security Area* addresses Internet security aspects.

- *Sub-IP Area* is devoted to the definition of technologies and protocols that normally are located at a layer below IP in the protocol stack and are devoted to the provision of services such as VPNs, traffic engineering, and transport of link layers or even circuit emulation.

- *Transport Area* is responsible for the definition of transport-related matters, such as QoS, transport-level protocols (for instance, recently transport protocols for carrying signaling was defined), and congestion control.

Each of these areas is led by one or two *area directors*. Area directors and the IETF chair are members of the Internet Engineering Steering Group (IESG), which has the role of standards quality evaluation and can strongly influence the transition of an Internet Draft to proposed standards RFC status, by returning it to the WG until it attains an adequate level of quality to be published. The following section explains this role in greater detail.

## IETF Documents and Standardization Process

The IETF standardization process is quite different from that of the 3GPP. First, no company can officially be an IETF member. IETF membership is only allowed for engineers and scientists or students interested in the evolution of the Internet. These individuals, however, are more often than not sponsored by companies and organization, whose interests are therefore indirectly represented.

Second, there is no actual formal document evaluation process. When an individual deems something is needed to add functionality to the Internet, he or she (possibly with multiple other coauthors) submits an Internet Draft to the relevant IETF Working Group. If no appropriate WG exists, interested individuals may set up one with IESG approval, going through a *bird-of-feather* (BoF) first round of discussion to gauge consensus on the need of the WG and its scope. Normally this takes place at an IETF meeting (there are three IETF meetings for each calendar year). It should be noted that once a WG is set up, individuals can summit Internet Drafts and discuss them on a WG mailing list. All the decisions are taken on the mailing list, based on evidence of "rough consensus" and some proof of having "running code" that testifies that the protocol being developed really works.

Once the WG is sufficiently happy with an Internet Draft evolved through amendments from the mailing list, the WG submits the draft to the IESG for their review. When the IESG has no further comments, the document is published as a Request for Comments (RFC) document. An RFC can be just an informational document, which documents something some group of individuals do or a protocol they use, or a standard track document, for instance, a protocol that is going to be generally used in the Internet.

In addition, there are different levels of standards track document. Initially, a standards track RFC is a *proposed standard*. Then, after some years of operational experience and with the evidence of at least two independent interoperable implementations, an RFC can become a *draft standard*. A draft standard RFC normally is a very stable document. After many years of operation, the IETF may elect to promote a draft standard RFC to the status of *Internet standard*. Other times, when the protocol becomes obsolete and no more widely used, the RFC can become "historical." Sometimes, if a WG or IESG needs to publish some rules or practices used in the Internet or in the IETF, they publish a *Best Current Practice* (BCP) RFC.

**NOTE** An individual may ask the IESG to evaluate directly a document he or she has produced, to document something that is relevant to the Internet operation. For instance, you might ask them to look at a proprietary protocol that happened to become widespread in the Internet, because it is supported by a dominant vendor of a popular internetworking device. The IESG may decide to approve recording the document as an informational RFC. Sometimes vendors misuse this option to advertise their own proprietary solutions as IETF standards.

## IEEE 802 LAN/MAN Standards Committee

The Institute of Electrical and Electronics Engineers (IEEE) defines standards for local area networking in the IEEE P802 LAN/MAN standards committee, part of the IEEE Standards Association (IEEE-SA). In recent years, the IEEE 802.11 WG has defined a standard for Wireless LAN, known as the 802.11 standard. This is a very promising set of standards, and it is creating a serious competitor (or a complementing technology, depending on the way the industry looks at it) to 3G technologies in serving network access in hot-spot areas such as airports, hotels, and train stations (more on this in Chapter 9). The IEEE 802.11 WG is organized into Task Groups (TGs). Each Task Group takes care of the standardization of a particular aspect of the WLAN technology, and they are the authors of the standards documents. Following is a list of 802.11 TGs, directly derived from the IEEE standards Web site:

**MAC Task Group.** [The scope of this project] is to develop one common MAC for Wireless Local Area Networks (WLANs) applications, in conjunction with the PHY Task Group work. Work has been completed on the ISO/IEC version of the original Standard, published as 8802-11: 1999 (ISO/IEC) (IEEE Std. 802.11, 1999 Edition).

**PHY Task Group.** The scope of the project is to develop three PHYs for Wireless Local Area Networks (WLANs) applications, using Infrared (IR), 2.4-GHz Frequency Hopping Spread Spectrum (FHSS), and 2.4-GHz Direct Sequence Spread Spectrum (DSSS), in conjunction with the one common MAC Task Group work. Work has been completed and is now part of the original Standard. Work has been completed on the ISO/IEC version of the original Standard, published as 8802-11: 1999 (ISO/IEC) (IEEE Std. 802.11, 1999 Edition).

**Task Group a.** The scope of the project is to develop a PHY to operate in the newly allocated UNII band. Work has been completed on the ISO/IEC version of the original Standard as an amendment, published as 8802-11: 1999 (E)/Amd 1: 2000 (ISO/IEC) (IEEE Std. 802.11a-1999 Edition).

**Task Group b.** The scope of the project is to develop a standard for a higher rate PHY in the 2.4-GHz band. Work has been completed and is now part of the Standard as an amendment, published as IEEE Std. 802.11b-1999.

**Task Group b-cor1.** The scope of this project is to correct deficiencies in the MIB definition of 802.11b. As the MIB is currently defined in 802.11b, it is not possible to compile an interoperable MIB. This project will correct the deficiencies in the MIB. It is an ongoing Task Group.

**Task Group c.** [This project] adds a subclause under 2.5 Support of the Internal Sub-Layer Service by specific MAC Procedures to cover bridge operation with IEEE 802.11 MACs. This supplement to ISO/IEC 10038 (IEEE 802.1D) will be developed by the 802.11 Working Group in cooperation with the IEEE 802.1 Working Group. Work has been completed and is now part of the ISO/IEC 10038 (IEEE 802.1D) Standard.

**Task Group d.** This supplement will define the physical layer requirements (channelization, hopping patterns, new values for current MIB attributes, and other requirements) to extend the operation of 802.11 WLANs to new regulatory domains (countries). It is an ongoing task.

**Task Group e.** This Task Group is expected to enhance the 802.11 Medium Access Control (MAC) to improve and manage Quality of Service, provide classes of service, and enhanced security and authentication mechanisms. [It will] consider efficiency enhancements in the areas of the Distributed Coordination Function (DCF) and Point Coordination Function (PCF). These enhancements, in combination with recent improvements in PHY capabilities from 802.11a and 802.11b, will increase overall system performance, and expand the application space for 802.11. Example applications include transport of voice, audio and video over 802.11 wireless networks, video conferencing, media stream distribution, enhanced security applications, and mobile and nomadic access applications. The security part of the TGe PAR (Project Authorization Request) was moved to the ongoing TGi PAR as of May 2001 [PARs are discussed in greater detail in the next section].

**Task Group f.** [This Task Group] develops recommended practices for an Inter-Access Point Protocol (IAPP), which provides the necessary capabilities to achieve multivendor Access Point interoperability across a Distribution System supporting IEEE P802.11 Wireless LAN Links. It is an ongoing TG.

**Task Group g.** The scope of this project is to develop a higher speed(s) PHY extension to the 802.11b standard. The new standard shall be compatible with the IEEE 802.11 MAC. The maximum PHY data rate targeted by this project shall be at least 20 Mbit/s. The new extension shall implement all mandatory portions of the IEEE 802.11b PHY standard. The current 802.11b standard already defines the basic rates of 1, 2, 5.5, and 11 Mbit/s. The proposed project targets further developing the provisions for enhanced data rate capability of 802.11b networks. It is an ongoing TG.

**Task Group h.** [The scope of this project is to enhance] the 802.11 Medium Access Control (MAC) standard and 802.11a High Speed Physical Layer (PHY) in the 5-GHz Band supplement to the standard; to add indoor and outdoor channel selection for 5-GHz license exempt bands in Europe; and to enhance channel energy measurement and reporting mechanisms to improve spectrum and transmit power management (per CEPT and subsequent EU committee or body ruling incorporating CEPT Recommendation ERC 99/23). It is an ongoing project.

**Task Group i.** [The scope of the project is to enhance the 802.11 Medium Access Control (MAC), thereby enhancing] security and authentication mechanisms. It is an ongoing project.

It is now possible to buy a WLAN PCMCIA card or an *access point* based on compliance with one or more of the documents authored by the task groups. Wireless Ethernet Compatibility Alliance (WECA) is an industry forum charged with the mission of certifying interoperability of IEEE 802.11 products. The WECA determines the criteria for compliance, based on references to the appropriate documents generated by IEEE 802.11 TGs.

## *IEEE Documents and Standardization Process*

The creation of a new IEEE standard happens via a Standards Project. This must be sponsored by a member of the IEEE SA Standardization Board. An IEEE Standards Project may be:

**New.** A document that does not replace or substantially modify another standard

**Revision.** A document that updates or replaces an existing IEEE standard

**Amendment.** An addendum or a substantive change to an exiting IEEE standard

**Corrigenda.** A document that contains only substantive corrections to an existing IEEE standard

Each project must be authorized by the board after a Project Authorization Request (PAR)—which defines the scope of the project—has been submitted. Once the project is approved, it has to generate a draft document, which will later undergo a ballot, before being approved as an IEEE standard by the Review Committee, or *Revcom*. The Revcom makes recommendations to the IEEE-SA Standards Board on the approval or disapproval of documents submitted to IEEE-SA Standards Board.

IEEE standards can be classified in four ways:

**Standards.** These documents specify mandatory requirements.

**Recommended practices**. These documents clarify procedures and positions preferred by IEEE.

**Guides**. These define a set of alternative approaches, but no strict recommendations are made.

**Trial-use documents**. Valid for up to 2 years, these documents may belong to any of the preceding categories.

Every 5 years an IEEE standard has to undergo a *reaffirmation* process to confirm its validity. The IEEE process is summarized in Figure 1.2.

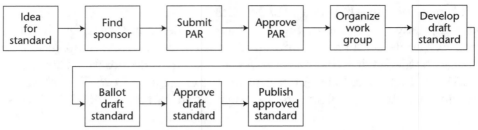

**Figure 1.2** The IEEE standards process.

## Finding Standards Documents Online

Throughout the book we frequently reference standards documents, and you are invited to probe further when you find a topic in which you are particularly interested. Fortunately, most of the documents are available via the Internet. Table 1.1 includes links to the Web sites where the standards documents we refer to can be downloaded.

**Table 1.1** Resources on the Web

| STANDARDS BODY | URL |
| --- | --- |
| 3GPP | http://www.3gpp.org |
| 3GPP2 | http://www.3gpp2.org |
| IETF | http://www.ietf.org |
| IEEE 802.11 WG | http://grouper.ieee.org/groups/802/11/ |

# Summary

In this chapter we introduced and defined what MVPN is, providing a market perspective that includes the customer drivers and key players, along with their needs for using MVPN technologies. We then stressed the importance of standards in MVPN solutions and provided a brief outline of the key standards bodies, specifying protocols and systems involved in MVPN technologies. In Chapters 2, 3, and 4, we provide a complete overview of the wireless systems and networking technologies required to understand MVPN technology. Readers who are already comfortable with data networking VPN technologies and wireless systems may want to skip ahead to Chapter 5.

# CHAPTER 7

# Data Networking Technologies

Before we enter the terminology and acronyms jungle that normally characterizes any telecommunications or data networking book, we need some tools to help us easily find the way out of the thickest parts of it and avoid becoming lost halfway through. This chapter equips you with a fundamental grasp of data networking technology required to understand the topics in this book. In the first section we introduce tunneling and labeling technologies that are fundamental to the provision of Virtual Private Networks over a shared internetworking infrastructure, as well as mobility support. We have chosen to handle IP security fundamentals in this section, since security is tightly coupled with tunneling technologies in VPN provisioning, and in fact one of the IP Security (IPSec) options (IPSec tunnel mode) is a tunneling mechanism frequently used in IP VPN service provisioning. In the second section we discuss quality of service (QoS) issues related to the provisioning of VPN services. This then takes us to the following section, which helps in understanding the importance of authentication, authorization, and accounting (AAA) to implement access control, a fundamental prerequisite for the ability to meet service levels and to charge for services. In the fourth section we address the necessity of network services to facilitate network operation and service provisioning.

After we complete this brief data networking tutorial, the next two chapters introduce the relevant aspects of cellular and Wireless LAN systems technology. Note that most of the topics in this chapter warrant their own book, and we will only scratch the surface. However, we will try to always reference standard documents and other worthwhile material for those interested in probing further.

## Tunneling and Labeling Technologies

Mobile VPN requires using technologies that leverage publicly available infrastructure, operated by service providers, that allows for "virtually private" connectivity between customer network sites and the mobile stations logically belonging to them, known as *Mobile VPN members* or *subscribers*. Such technologies (covered in depth in Chapter 5) are based on the encapsulation of the customer network data (also known as user data) packets into other packets, delivered using the networking technology of the shared network. This allows the use of the addressing scheme and the technology of the shared network, while delivering customer data belonging to networks that may be using different addressing schemes and different network or link layer protocols.

This encapsulation, or *tunneling*, as it is more often referred to in the data networking world, not only provides the ability to deliver data to and from mobile stations, but sometimes also adds integrity and confidentiality protection. Also, when the operator wants to support QoS, these technologies facilitate the delivery of predictable network transit, for instance, via traffic-engineered paths identified by a sequence of labels, like in Multi-Protocol Label Switching (MPLS). MPLS also provides the means to maintain the connectivity among multiple sites of a customer network in a fairly automatic way (like BGPv4/MPLS-based VPNs; [RFC2547]). Again, these topics are covered in greater depth in Chapter 5.

Sometimes the services offered by a carrier may simply be the forwarding of data from a wireless access gateway to the customer network site via a tunnel or a fixed access line. (See Chapters 6 and 7 for the definition of the wireless gateway entities in 3GPP- and 3GPP2-compliant systems.) Other times the service may extend to a managed multiple-sites VPN service, where the wireless access gateway becomes simply one of the customer network sites. Tunnels are also used to support mobility, by keeping one endpoint fixed and having the other "follow" the mobile data node at its point of attachment to the network (where normally the link layer of the access network is terminated). Mobile IP and GPRS Tunneling Protocol (GTP), covered in this chapter, are good examples of the latter.

The data may be transferred at the network layer or at the link layer using a protocol such as Point-to-Point Protocol (PPP). In this case, the wireless network simply terminates the wireless access protocols and relays the PPP or other link layer protocol to a network access server in the customer network. This is pretty much always the case with circuit-switched data-based MVPNs, but it can also be frequently encountered in packet-based wireless data services, where it might be favored because of the wide use of PPP-based remote access by enterprises on wireline access media. The approach involving PPP is normally based on a tunneling protocol called L2TP.

## Layer Two Tunneling Protocol

Layer Two Tunneling Protocol (L2TP) is defined in [RFC2661] as an IETF protocol that provides a standard approach to the tunneling of PPP frames over IP. Cisco and Microsoft had originally developed proprietary ways to accomplish this (via Layer Two Forwarding, or L2F, and Point-to-Point Tunneling Protocol, or PPTP, respectively), but the industry recognized the need for a standard-based approach. As a result, the IETF PPP Extensions (PPPEXT) Working Group was chartered with the task to define such a standard. The outcome was a tunneling protocol that could potentially be transported on any cell, frame, or packet-based transport network. In particular, the widely used UDP/IP transport was chosen as the preferred protocol (UDP stands for User Datagram Protocol).

L2TP defines two network entities with two distinct roles to be the peers for this protocol:

- The L2TP Access Concentrator (LAC) is located at the point of termination of access network protocol, and it can establish tunnels toward appropriate L2TP network access servers (LNSs).

- The LNS terminates tunnels from LACs and also offers network access services such as user authentication and address assignment.

An *LAC client* running on a laptop or any other suitable computing device could also be used to initiate L2TP tunnels toward LNS. The LAC client-based usage of L2TP constitutes a technology-independent way to access remote networks, provided that these are reachable at the network layer. Figure 2.1 illustrates the model we have just discussed.

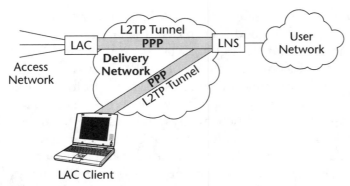

**Figure 2.1**   The L2TP model.

L2TP defines a reliable control channel. Over this control channel it is possible to establish a tunnel between the LAC and the LNS. The tunnel establishment phase normally includes authentication via the L2TP exchange of a secret between the LAC and LNS (in the form L2TP tunnel; password). The authentication of the party attempting to set up a tunnel is important, since it is not desirable to have an LNS accept any L2TP commands coming from an unknown LAC if they are not authorized to do so. However, since the L2TP protocol does not come with data origin authentication and confidentiality, L2TP cannot be considered a secure protocol. In fact, it is still possible for an attacker to send packets to an LNS or an LAC and impersonate each node's peer. Securing L2TP requires another IETF protocol suite defined for the support of IP security: IPSec. We provide details on IPSec and IPSec modes later in the chapter, adding more information and explaining how to secure L2TP tunnels.

When a tunnel is established between the LAC and LNS, it is possible to set up and tear down a PPP session and to forward associated frames between the two nodes using the L2TP encapsulation format over the L2TP data channel. The L2TP header (see Figure 2.2) includes the Tunnel ID and the Session ID information to enable two levels of multiplexing. The Tunnel ID defines a tunnel between two peers, and it therefore implicitly identifies the peer node at the receiving end. The Session ID identifies the particular PPP session within the tunnel. Because the Session ID information can be exchanged only after the tunnel between the LAC and LNS is in place, PPP call setup latency could be reduced if the L2TP tunnel is already set up when a PPP session needs to be handled. Often carrier-class deployments establish L2TP tunnels up front.

```
0                   1                   2                   3
0 1 2 3 4 5 6 7 8 9 0 1 2 3 4 5 6 7 8 9 0 1 2 3 4 5 6 7 8 9 0 1
```

| T|L|x|x|S|x|O|P|x|x|x|x|   Ver | Length (opt) |
|---|---|
| Tunnel ID | Session ID |
| Ns (opt) | Nr (opt) |
| Offset Size (opt) | Offset pad... (opt) |

**Figure 2.2** The L2TP message header.

User authentication within PPP sessions normally takes place transparently to the LAC. The LAC merely decides to which LNS the L2TP session has to be set up and subsequently forwards the incoming PPP frames to it. The selection of the LNS can be based on information such as the destination number called, or when used in GPRS network, on the identifier of the network the PPP user is requesting access to. Forwarding PPP frames to the correct LNS lets the PPP authentication phase occur between the LNS and the PPP client on the remote device. The LAC can also perform *proxy* authentication by collecting authentication data from the incoming call and relaying it to the LNS using L2TP signaling. This requires mutual trust relationships to be established between the LAC and LNS operators. The LNS, after it has received the proxy authentication data from the LAC, may later optionally reauthenticate the user at the PPP level by initiating a new PPP authentication phase before moving to the configuration of the network layer.

The LAC may determine the LNS IP address dynamically based on the received username and password, which may contain the Network Access Identifier (NAI), defined in [RFC2486]. The LAC can in this case conduct a first pass of user authentication with the AAA infrastructure (see the "Authentication, Authorization, and Accounting" section later on page 63). The AAA infrastructure determines the user Home AAA server based on the NAI domain component. The AAA infrastructure could return, when the user is granted access, L2TP tunnel information such as the LNS IP address and the L2TP password. In fact, the LAC may decide whether the user requires an L2TP tunnel to an LNS or simply access to a network directly attached based on the domain component of the username (formatted like this: domain\user). In this way, it works like a regular network access server (NAS). For instance, user JDoe, may want to access the Internet using the username *Inet-access\JDoe*, and the corporate network using L2TP via the username *Corpnet-access\JDoe*. L2TP can handle both calls coming from the access network to the LAC, denominated "incoming calls," as well as requests from the LNS to call a specific terminal on the access network (to implement, for example, a call-back service), denominated "outgoing calls."

Given its flexibility and its rich set of options, L2TP has become widely used to "divorce" the location of access termination from the location of termination of the PPP protocol, with large deployments in global remote access facilities for large corporations (see Chapter 5). It became a de facto standard for services such as remote access outsourcing when an enterprise relies on a service provider to handle their remote worker's PPP sessions at their facilities (POPs equipped with remote access servers) and then relay them to corporate data center for authentication and IP address assignment. Given current L2TP popularity with corporations, it is not surprising that both the GPRS/UMTS and CDMA2000 standards allow for its use as a way to support compulsory access to corporate networks (more on this in Chapters 6 and 7), thus providing an easy way to integrate wireless and wireline access methods. Further information on L2TP can be obtained from [Black2000].

## IP in IP Tunneling

*IP in IP*, also referred to as IPIP, is the most basic tunneling service; it encapsulates an IP packet into another IP packet. This encapsulation method is specified in [RFC2003], which has been developed as a companion document to [RFC2002] (the original Mobile IPv4 specification). In IPIP the outer IP packet header identifies the addresses of the tunnel endpoints, where the source address is the address of the encapsulator and the destination address is the address of the decapsulator.

In recognition of the fact that sometimes encapsulating an IP packet in another IP packet may lead to excessive overhead, especially when small payload IP packets are tunneled, it was necessary to define a way to compress the information associated with the inner IP packet header. [RFC2004] describes the minimal IP in IP encapsulation that defines an encapsulation header inserted between the outer IP packet and the inner packet payload so that the decapsulator can reconstruct the inner IP packet header (see Figure 2.3). This can lead to 8 to 12 bytes of saving per packet.

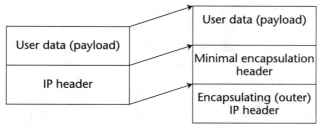

**Figure 2.3** The minimal encapsulation for IP.

**NOTE** Both IP in IP and minimal encapsulation for IP tunneling protocols rely on other protocols (e.g., Mobile IP) or network element provisioning to get the tunnel set up. Also, IP in IP by itself is not secure and requires IPSec for this function. This combination is normally referred to as *IPSec tunnel mode.*

## GRE Protocol

The Generic Routing Encapsulation (GRE), specified in [RFC2784], is an IETF standard defining multi-protocol encapsulation format that could be suitable to tunnel *any* network layer protocol over *any* network layer protocol. This concept was originally specified in [RFC 1701], which was an informational RFC. When this original protocol was moved to a standards track, the decision was made to replace it with two separate RFCs: [RFC2784] and [RFC2890]. [RFC2890] is an extension of the basic GRE header described in [RFC2784]. It was determined necessary because [RFC 2784] does not lend itself to encapsulation of PPP frames, since it does not have a sequence number in the GRE encapsulation format. This limitation was removed by adding a *sequence number* extension to the basic GRE header. Also, [RFC2784] does not allow for multiplexing onto the same GRE tunnel of tunneled packets belonging to different administrative entities possibly adopting overlapping private address spaces (a very useful feature for the provision of Virtual Private Networks). This limitation was also removed by adding a *key field*—that is, a numeric value used to uniquely identify a logically correlated flow of packets within the GRE tunnel—as an extension of the basic GRE header. These extensions to a basic GRE defined by [RFC2890] were especially useful in wireless data communications. For example, they allowed for in-sequence delivery of

PPP frames over the R-P interface in CDMA2000 (see Chapters 4 and 7 for more details), and the provisioning of compulsory MVPN services.

GRE, as defined by these RFCs, is normally used in two classes of applications: the transport of different protocols between IP networks and the provision of VPN services for networks configured with potentially overlapping private address space. The GRE header key field can be used to discriminate the identity of the customer network where encapsulated packets originate. In this way, it provides a way to offer many virtual interfaces to customer networks on a single GRE tunnel endpoint. This feature allows for policy-based routing (that is, when routing decisions are not based only on the destination IP address but on the combination of a virtual interface identifier, and the destination IP address) and relatively easy per-user network accounting. Also, a GRE header allows the identification of the type of the protocol that is being carried over the GRE tunnel, thus allowing IP networks to serve as a bearer service onto which a virtual multi-protocol network can be defined and implemented.

**NOTE** Similar to the IP in IP tunneling mechanism, the GRE tunneling technology does not include a tunnel setup protocol. It requires other protocols, such as Mobile IP, or network management to set up the tunnels. It also does not include security mechanisms and must be combined with IPSec to support secure user data delivery.

## Mobile IP

Mobile IP can be based on a variety of tunneling mechanisms, but by itself, it does not provide one. As we mentioned in the "IP in IP Tunneling" section, Mobile IP was originally defined by [RFC2002], which later was made obsolete by the more up-to-date [RFC3220] to support host mobility at the network layer (IP). Mobile IP was originally conceived to let nodes with fixed static IP addresses to be permanently reachable even if they changed their point of attachment to the network. Its applicability was limited to supporting the mobility of Internet hosts and routers only in certain environments, such as campus or university networks, or for defense applications. There was no support for dynamic address assignment and no accounting and user authentication features. However, Mobile IP was later adopted by cellular systems such as Motorola IDEN deployed by Nextel (a United States based operator) and more recently by CDMA2000 standard for core networks, and for these systems, it evolved to take into account the needs of commercial environments. Today Mobile IP is being increasingly considered a preferred method for support of multi-access technologies and inter-system roaming [Solomon1998].

Mobile IP is defined as a technology-independent tunnel setup and maintenance protocol used to allow roaming Internet hosts or routers to maintain constant IP address and uninterrupted IP-level connectivity back to a home network while changing points of attachment to a network and the network access technology.

**NOTE** Throughout the book, we will interchangeably use the terms mobile node (MN) in accordance with Mobile IP and IETF terminology, or mobile station (MS) in accordance with 3GPP terminology.

### *Implementing Mobile IP*

Like L2TP, Mobile IP provides an architectural model that defines the roles of different entities that may be involved in Mobile IP operation. Mobile IP must be implemented in three main functions:

- Home Agent (HA)
- Foreign Agent (FA)
- Mobile Node (MN)

HA and FA support the Mobile IP protocol on the mobile node's home and foreign (visited), networks respectively.

During the mobile IP communication session the HA is being continuously informed by the MN of its current location via Mobile IP Registration Request messages, as the MN roams through different networks. The HA and FAs available in Foreign Networks advertise their availability by means of *agent advertisement* messages broadcasted over the links directly attached to them. The location of the MN is represented by its *care-of address* (CoA) being temporarily assigned to it by the FA (or acquired by the MN itself in the visited network when Mobile IP operates in collocated care-of address mode). The HA forwards the traffic to MN (almost always through a FA) and accepts it from MN (possibly via FA) when reverse tunneling is used (see next paragraph). It also processes FA registration requests and manages Mobile IP tunnels for users gaining network access via the HA itself. The typical Mobile IP network model is shown in Figure 2.4.

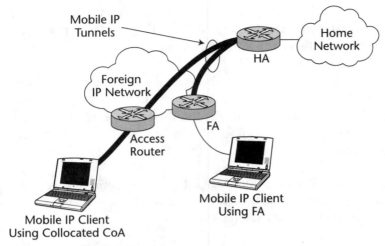

**Figure 2.4**   Mobile IP model.

The FA may forward packets from the MN directly to the Internet or tunnel them back to the HA in what is called Mobile IP *reverse tunneling*. Reverse tunneling is often necessary to avoid ingress filtering to drop packets from the MN, as well as to let corporations and network access providers operating the HA create VPNs in which all the traffic to and from the MN can be optionally secured and would always traverse the HA in its private network for tighter control and security. The standards also allow the MN to directly register with the HA when FA is not available, after it acquires an IP care-of address in the visited network, which in this case is called a *collocated care-of address*. This address would then be used by the MN to exchange Mobile IP registration messages and to tunnel packets to and from the HA, all without the need of an FA. Interestingly, this last feature did not find its way into the Telecommunications Industries Association (TIA) specification for CDMA2000, partially under the influence of wireless operators who would potentially lose control over the mobile users, since the mobile users could at will bypass the operator-owned FAs if this option were implemented.

An MN can be assigned a static IP address or it can obtain it dynamically from an IP address pool belonging to networks served by the HA. The FA, being a router in the visited network, is capable of serving visitor MNs. As depicted in Figure 2.5, the FA advertises its presence via periodical Mobile IP agent advertisements, or by responding with Mobile IP agent advertisements to Mobile IP *agent solicitation* messages sent by MNs visiting one of the networks served by the FA. When an MN moves to a network controlled by a certain FA, it can send a Mobile IP *registration request* (RRQ) to the FA, which in turn registers the MN with its HA by sending an RRQ to the appropriate HA. The HA may accept this RRQ (after authenticating the sender) and issue a Mobile IP *registration response* (RRP) to the FA, which in turn relays it to the MN. After this message exchange and all authentication phases are complete, the HA creates the IPIP or GRE tunnels to the FA and then start forwarding packets destined to a registered MN to the FA, which then relays them to the MN.

The Mobile IP protocol itself has undergone many changes and additions over the last decade, often referred to as *extensions*, to make it suitable for commercial deployment in various systems, especially because of the requirements of TIA specified in IS835, which defines Mobile IP as a basis for network layer user data mobility in CDMA2000.[1] For instance, replay protection has been added to the Mobile IP registration messages by the use of a Challenge-Response extension. Also, roaming support via the NAI extension (see later in the chapter for a description of NAI-based roaming) and dynamic home address assignment were added.

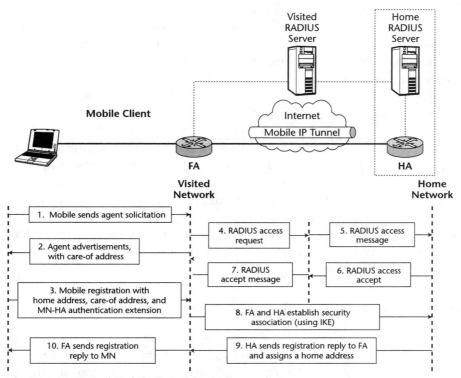

**Figure 2.5** Typical Mobile IP registration procedure.

---

[1] See Appendix A, which lists the details of the majority of Mobile IP extensions.

Dynamic home address assignment is an important Mobile IP feature which deserves a little more attention. Originally, Mobile IP was based on the assumption that all HAs in the network were statically allocated. This model, however, is less suitable for commercial wide-area deployment because it exposes service providers to extremely inefficient use of IP addresses. This model also makes HA a single point of failure, because once the HA serving a particular MN fails, the MN is denied the data services until this particular HA is back in service. Along with addressing these problems, dynamic HA allocation would provide more optimal routing and better utilization of landline data infrastructure when a MN is at a significant distance from its home network. This is why the second release of IS835 standards will include the ability to dynamically allocate the HA. More information on this topic as well as the examples of commercial Mobile IP deployment within CDMA2000 system framework can be found in Chapters 4 and 7.

The standards are also currently working on the definition of the so-called *optimal routing*. In this definition, the MN may update the node involved in a session, named the correspondent node (CN), of the IP address it has acquired in the visited network (the collocated CoA or the FA-generated CoA). Then the CN would be requested to send packets directly to this IP address. Optimal routing, however, proved not to scale well in a secure way, since the burden for correspondent nodes of millions of mobile nodes (such as Web servers) would be quite substantial. Other operational aspects of optimal routing are also not yet clear, which makes this a mode of operation with many open issues and question marks.

In contrast to IPv4, Mobile IP for IPv6 was strongly based on optimal routing and was expected to be supported by any IPV6-compliant host or router. This proposition, however, is currently encountering stumbling blocks in its standardization process because of security concerns about its reliance on trust in extension header processing and nonscalability with regard to security. Therefore, Mobile IPv6 deployment cannot be expected in its current shape until changes are provided and ratified by Mobile IP WG to make it deployable.

## GPRS Tunneling Protocol

GPRS Tunneling Protocol (GTP) as originally defined in [3GPP TS 29.060] and [GSM TS 09.60], is a protocol used to support mobility of GPRS and UMTS mobile stations (MSs) roaming among geographical locations served by different *Serving GPRS Support Nodes* (SGSNs). The MS home network is represented by *Gateway GPRS Support Nodes* (GGSNs). GGSNs connected to SGSNs via GTP. As opposed to Mobile IP, GTP does not have to be supported in the MS. It is a protocol used only within the network, and the GTP protocol interworks with other protocols in order to interact with the MS and thus allow for tracking of its current whereabouts. In this section we do not present the details of the application of GTP in the GPRS and UMTS systems, since we discuss this extensively in Chapters 4 and 6. Instead, we focus on the description of the protocol and its technical details.

There are two versions of the GTP protocol. Version 0, described in [GSM TS 09.60], is used in the systems based on the GSM Base Station Subsystem (BSS) and applies to GSM Release 97 and 98. Version 1 is described in [3GPP TS 29.060] and applies to systems based on both GSM and UMTS radio access networks. 3GPP decided to create a new version of GTP protocol—not backward-compatible with GTPv0—because the group members wanted to introduce new features that could not be supported using the older version of the GTP protocol. They wanted to enable the separation of the protocol in a user plane protocol (GTP-U) and a control plane protocol (GTP-C). The reason for this split was dictated by the need to support GTP-U-based tunneling of user data over the interface between the UMTS core and the UMTS radio access network (Iu interface), without the need to use GTP-C to set up Iu tunnels. Another significant feature that differentiates GTPv1 from v0 is the support of multiple QoS levels per IP address assigned to an MS, which requires the establishment of multiple UMTS bearers and the use of multiple tunnels per MS data session. This led to the need of a multiplexing field in the GTPv1 header (the Tunnel Endpoint Identifier), which was then also used to replace a quite cumbersome and complex structure of the fields identifying a data session in GTPv0. Figure 2.6 illustrates the difference between GTPv0 and v1 header structure.

GTPv0 could be transported over TCP or UDP. TCP was supposed to be used to offer reliable transfer of user data, which would be required for the transport of X.25 data. The wireless data networking market, though, by abandoning X.25 option has evolved in a way that made the use of TCP as a transport protocol for GTP redundant, and starting from UMTS R'99, GTP will be transported only over UDP. The UDP port number for GTPv0 is 3386, and the port numbers for GTPv1 are 2023 for GTP-C and 2052 for GTP-U.

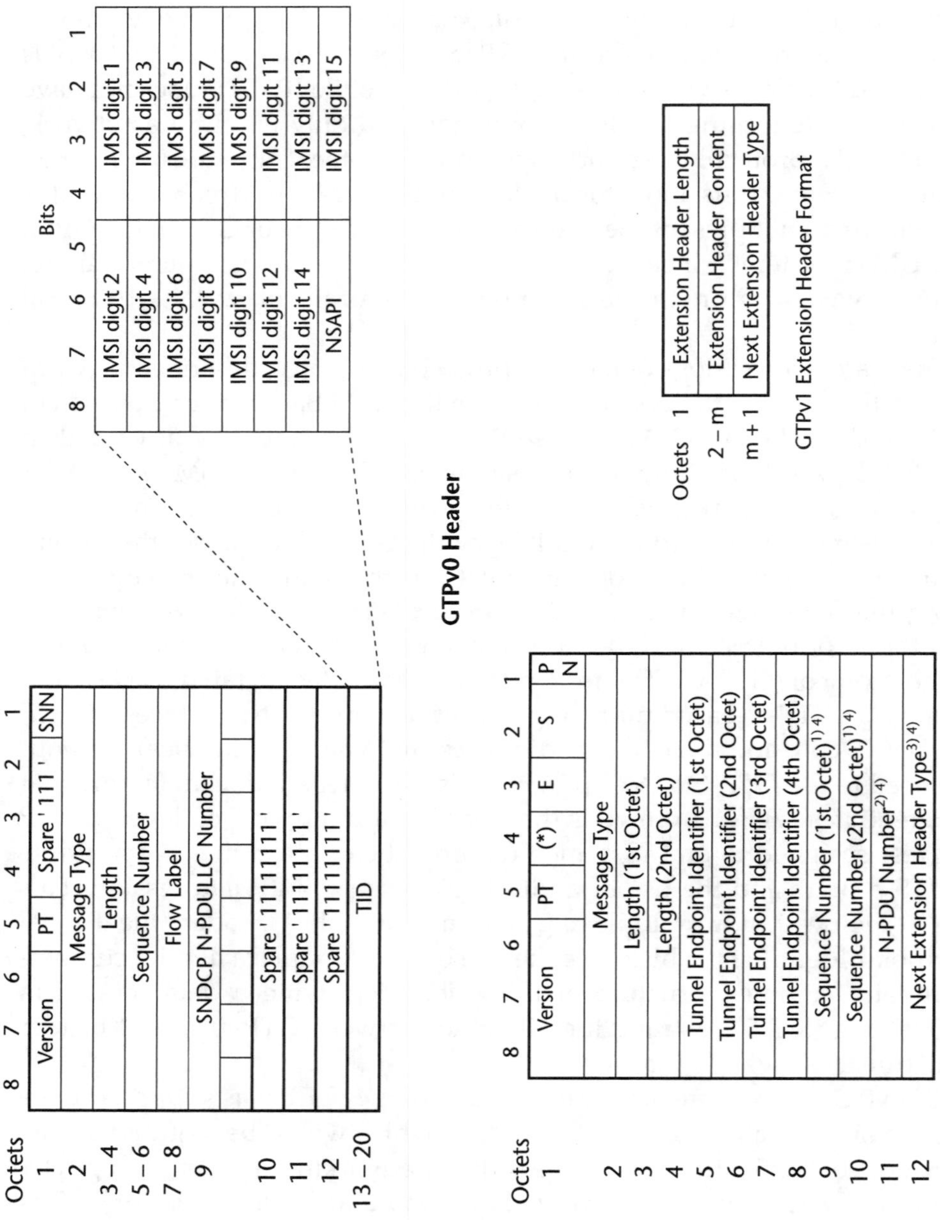

**GTPv0 Header**

| Octets | Bits | | | | | | | |
|---|---|---|---|---|---|---|---|---|
| | 8 | 7 | 6 | 5 | 4 | 3 | 2 | 1 |
| 1 | Version | | PT | | Spare ' 111 ' | | | SNN |
| 2 | Message Type | | | | | | | |
| 3 – 4 | Length | | | | | | | |
| 5 – 6 | Sequence Number | | | | | | | |
| 7 – 8 | Flow Label | | | | | | | |
| 9 | SNDCP N-PDULLC Number | | | | | | | |
| 10 | Spare ' 11111111 ' | | | | | | | |
| 11 | Spare ' 11111111 ' | | | | | | | |
| 12 | Spare ' 11111111 ' | | | | | | | |
| 13 – 20 | TID | | | | | | | |

| | 8 | 7 | 6 | 5 | 4 | 3 | 2 | 1 |
|---|---|---|---|---|---|---|---|---|
| | IMSI digit 2 | | | | IMSI digit 1 | | | |
| | IMSI digit 4 | | | | IMSI digit 3 | | | |
| | IMSI digit 6 | | | | IMSI digit 5 | | | |
| | IMSI digit 8 | | | | IMSI digit 7 | | | |
| | IMSI digit 10 | | | | IMSI digit 9 | | | |
| | IMSI digit 12 | | | | IMSI digit 11 | | | |
| | IMSI digit 14 | | | | IMSI digit 13 | | | |
| | NSAPI | | | | IMSI digit 15 | | | |

| Octets | | | | | | | | | |
|---|---|---|---|---|---|---|---|---|---|
| | 8 | 7 | 6 | 5 | 4 | 3 | 2 | 1 |
| 1 | Version | | PT | (*) | | E | S | P N |
| 2 | Message Type | | | | | | | |
| 3 | Length (1st Octet) | | | | | | | |
| 4 | Length (2nd Octet) | | | | | | | |
| 5 | Tunnel Endpoint Identifier (1st Octet) | | | | | | | |
| 6 | Tunnel Endpoint Identifier (2nd Octet) | | | | | | | |
| 7 | Tunnel Endpoint Identifier (3rd Octet) | | | | | | | |
| 8 | Tunnel Endpoint Identifier (4th Octet) | | | | | | | |
| 9 | Sequence Number (1st Octet)[1) 4)] | | | | | | | |
| 10 | Sequence Number (2nd Octet)[1) 4)] | | | | | | | |
| 11 | N-PDU Number[2) 4)] | | | | | | | |
| 12 | Next Extension Header Type[3) 4)] | | | | | | | |

| Octets | |
|---|---|
| 1 | Extension Header Length |
| 2 – m | Extension Header Content |
| m + 1 | Next Extension Header Type |

GTPv1 Extension Header Format

**Figure 2.6**  GTPv0 and GTPv1 headers.

GTP-C messaging includes the following:

- Tunnel management messages used to detect fault conditions, loss of connectivity, and peer node restart.

- Session management messages used to set up tunnels between GGSNs and SGSNs, and to update the per-node changes in the tunnel parameters such as QoS, new user plane, and control plane IP addresses of the new SGSN the MS has moved to.

- Location management messages used to implement network-initiated GTP session setup procedures.

- Mobility management messages used to transfer MS context and session context information at handoff time.

GTP-U is simply used to encapsulate user packets, but it can also monitor the transmission path for failures using tunnel management messages. It is used between SGSNs, between GGSNs and SGSNs, and between the UMTS SGSN and the UMTS RNC.

When a session is set up, GTP can transfer MS and subscriber-related data to the GGSN—using the "Protocol Configuration Options" *Information Element* (IE)—as well as some information on whether the subscriber has access rights, via subscription, to the *access point* (AP) network where the GTP tunnel is about to be created. This is performed via the Selection Mode IE, which transfers to the GGSN information on whether the AP was subscribed to by the user or just selected without subscription. This information is inserted into GTP by the SGSN. The IMSI (a globally unique identifier of the subscriber) and the MSISDN (the MS phone number) are sent by the SGSN to the GGSN using GTP-C, and the GGSN can relay this information to external servers, which, for instance, could apply user-identity-based policies. Also, GTP is used between SGSNs to transfer MS-related information at handoff time. It is also used when the MS attaches to an SGSN, and this new SGSN needs to retrieve subscriber identity information from the SGSN the MS was previously attached to before the MS last powered off (or, better, performed a detach procedure). More information on GPRS and UMTS mobility can be found in [Eberspacher2001] and [Kaaranen2001].

GTP can be used to encapsulate different user data protocols such as PPP, IPv4, and IPv6. Other protocols such as X.25 were originally allowed by the standard, but later the support for them was dropped by both vendor and operator communities along the standard evolution and maintenance path. It must be mentioned that GTP by itself is not secure and requires IPSec to add integrity protection and confidentiality. It does not require other protocols for GTP tunnel setup, however. GTP-U tunnels can be established by GTP-C or other protocols, such as Radio Access Network Application Part (RANAP) in UMTS.

## Addressing Security

There is a common perception that IP is not a secure protocol and that the public Internet is exposed to all sorts of attacks from all sorts of individuals and groups, from the bored teenager who hacks Internet hosts to cybercriminals using the Internet to damage institutions or break into banks. Indeed, this perception is correct. Being insecure is a recognized weakness of IP technology. Data integrity, data origin authentication, and confidentiality need to be provided either by:

- Adding security mechanisms to the applications and devices using IP (examples include Web-based services such as e-commerce, e-banking, or access to corporate mail via Web interfaces).

- Securing the IP protocol itself via some extensions and protocols defined by the IETF. These IP protocol extensions allow for an adequate level of security and are known under the umbrella name IPSec. IPSec is the name of the IETF Working Group that has developed an *Internet security architecture* [RFC2401] and the extensions to the IP protocol that this architecture has incorporated.

In this section, we describe both kinds of approaches—that is, network layer security provided via IPSec and application layer security provided by Transport Level Security (TLS, which recently superceded SSL). We also discuss security infrastructure enablers such as the Public Key Infrastructure (PKI), which supports the secure exchange of information and identity verification over the Internet. More information on both IPSec and PKI can also be obtained from multiple books such as [Doraswamy1999].

### IPSec

The IPSec architecture defines the components necessary to provide secure communication between IP protocol peer entities, along with the related terminology. IPSec extends the IP protocol with two extension headers: the ESP header (the IP Encapsulating Security Payload, defined by [RFC2406] and the AH (Authentication Header, defined by [RFC2402]). The ESP is used to provide implicitly data confidentiality, payload integrity, and authentication, whereas AH is used to offer payload data integrity and guarantees the integrity of the nonmutable fields of the IP header as well. Both of these headers can be used either to encapsulate an IP packet in another IP packet (IPSec tunnel mode) or to encapsulate only the payload of an IP packet (IPSec transport mode). In Figure 2.7, AH is used to provide IPSec transport mode and ESP to provide IPSec transport mode, but a combination of AH and ESP is also possible, according to the standards.

Although interoperable implementations of AH exist, in the VPN industry the ESP tunnel and transport modes are the most commonly used approaches. This is because the AH only provides the subset of ESP capabilities and because, by including in the authentication algorithm all the nonmutable IP header fields, the data origin authentication provided by AH can be offered by using IP tunnel mode with ESP. In fact, with the encryption service offered by the ESP tunnel mode, the inner IP packet, IP header, and payloads are implicitly protected from alteration along the route from tunnel ingress point to tunnel egress point. AH is nevertheless used by some protocols, such as Mobile IP, which require control messages to be protected via AH transport mode (and their encryption is optional). These security mechanisms, however, are general and are not forcing the use of a predefined encryption or authentication algorithm. Therefore, implementations can add encryption algorithms as they become available, without changing the architectural model. The most commonly used encryption protocol is Triple Data Encryption Standard (3DES), and the most commonly used authentication protocols are based on hash functions such as SHA-1 and MD-5. (SHA stands for Secure Hash Algorithm, MD for Message Digest.)

Fundamental components in the IPSec architecture are the Security Policy Database (SPD) and Security Association Database (SAD). Every IP interface for which IPSec is enabled must be equipped with a database of security classification rules and security actions. Each individual rule and action pair is known as a *security policy*. A *security association* (SA) defines a unidirectional packet treatment in terms of security policy enforcement actions that define which IPSec headers are applied, which encryption or authentication algorithms are used, and which keys are used to execute these algorithms. For each IP interface, there is a pair of such databases: one for the inbound traffic and one for the outbound traffic. If a packet does not match any rule, the interface may be configured to discard it.

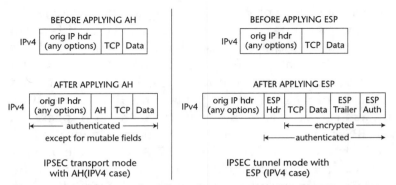

**Figure 2.7**  IPSec tunnel mode and transport mode with ESP and AH.

To better understand these concepts, we can use the following example of an entry in an outbound IPSec SPD and SAD. A possible security policy can be defined by the following entry in the SPD of an IP interface:

"For all packets bound for destination IP address (192.43.56.82) and port number 8080, apply security association *ALFA.*"

Security association *ALFA* is an entry in the SAD of the same IP interface, defined as based on IPSec tunnel mode with ESP and encryption algorithm 3DES and with an encryption key manually exchanged and provisioned at the endpoints. In the literature this SA is known as a *symmetric key—based* SA.

Security keys can be symmetric or asymmetric. Symmetric, or private, keys are distributed to both parties involved in a secure communication. Asymmetric keys are based on the RSA Data Security Inc. patented *public keys* cryptography paradigm, widely used in the industry to perform both encryption and authentication. In this setup, one party that wishes to engage in secure communications with others makes available a public key for retrieval at a well-known public keys repository. This approach is known as *asymmetric* key based because it uses a pair of keys: one that is public and widely distributed and another that is kept secret and never disclosed. Material encrypted using a public key can be decrypted only by using the associated private key. Conversely, only the public key can be used to decrypt material encrypted using the private key.

An asymmetric key system can be used to exchange a secret key necessary to run a symmetric-keys-based encryption algorithm. In other words, if a party knows the public key of an entity, it can send it a secret key, encrypted using the public key, and this party could unencrypt it using the private key and further use it for a symmetric-key-based encryption communication. To communicate with a peer using a public key, it is necessary to trust the source of this key. It is therefore necessary that the repositories of such information can be trusted (for instance, their public key is known and they digitally sign the public keys they hand out using their private key). These repositories are known as certificate authorities (CAs), and they form the base of the PKI. CAs and the PKI are discussed in greater detail in the next section.

An SA can be manually provisioned or dynamically managed, together with the security keys necessary to run the encryption or authentication protocols. This protocol is known as SA and Key Management Protocol, and the current IETF standard for this is known as Internet Key Exchange (IKE), [RFC2409]. Over time, IKE has undergone criticism of some of the engineering choices in its design. The IPSec Working Group in the IETF is currently mulling its evolution, and we can expect with some likelihood an IKEv2 at some time to come.

The IPSec protocol can be deployed in host-to-host, host-to-router, or router-to-router form. A router implementing IPSec and applying security policies to IP traffic is often referred to as an *IPSec gateway*. Figure 2.8 exemplifies the host-to-router and router-to-router cases, which are of special interest for VPN service provisioning.

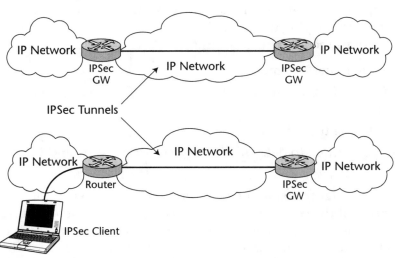

**Figure 2.8**  Sample IPSec architecture (gateways and hosts).

When IPSec is used for site-to-site IP VPNs (Virtual Private Networks built by connecting different sites of a corporate network using a shared IP-based infrastructure, see Chapter 5 for further details), IPSec tunnel mode is used. An example of IPSec transport mode application is L2TP tunnels integrity and confidentiality protection. It makes sense to use transport mode for L2TP tunnels protection because L2TP is a tunneling technology itself and the use of IPSec tunnel mode would result in redundant encapsulation, or nested tunnels. On the contrary, it makes much sense to use IPSec tunnel mode by itself for site-to-site VPNs, since packets need to be tunneled anyway and this mode provides both the tunneling and security required for this application.

## Public Key Infrastructure

Public Key Infrastructure is an important security concept and needs to be carefully defined. Public keys, briefly described in the previous section and used in data networking to verify digital signatures, by themselves do not carry any information about entities providing the signatures. The data networking industry recognized this problem and adopted security certificates binding the public key and the identity of the key-issuing entity, which in turn can be verified using a trusted public key, perhaps known using a certificate issued by a higher hierarchical level authority. The certificates are issued and enforced by a certificate authority, which is authorized to provide such services to positively identified entities requesting them. To perform its functions, a CA must be trusted by all entities (PKI members) relying on its services.

A *certificate* contains the following information:

- The certificate issuer's name
- The entity for which the certificate is being issued (also known as the subject)
- The public key of the subject
- Time stamps (needed to determine the age of a certificate and its validity)

All certificates are signed using the CA's private key. A user of the certificate can verify that the certificate information is valid by unencrypting the signature and verifying it matches the digest of the content received in the certificate. The signature is normally an encrypted digest of the content—that is, a string of bits obtained by encrypting the result of some nonreversible elaboration of the content.

PKI members may agree on a standard lifetime of a certificate and thus determine when a certificate is stale or expired. Also, a certificate authority can publish a certificate revocation list (CRL), so that the PKI members can consult certificates no longer validated by the CA by checking them against the CRL.

The trust relationships between the CA and other PKI members must be established prior to any PKI transactions. Such relationships are usually outside the scope of PKI and for that matter outside the scope of networking technology. PKI trust relationships can be established on geographical, political, sociological, business, or ethnic basis and can span industries, countries, population groups, or other entities bound by common interests. The PKI trust models could in theory be based on a single CA, which would be used to create a worldwide PKI similar to the worldwide Internet or a hierarchy of distributed CAs (see Figure 2.9) in which every CA can be trusted by following a certificate chain to a common certificate authority the parties entering secure communication trust.

In Figure 2.9 we describe the case of two parties, A and B, willing to exchange a secret. Party A retrieves B's public key from B's certificate. The certificate can be verified because it is signed with B's CA private key, and this can be checked by retrieving B's CA public key from B's CA certificate, which in turn can be verified by using the public key of a root certification authority that is guaranteed to be valid—for instance, because it is burned into the code of the PKI client on A's software module. Once B's public key is available, A encrypts the secret using it, and then it can send this encrypted message to B, along with its own (that is, A's) certificate and a digest of the encrypted secret, computed using A's private key. On reception of this message, B checks that the encrypted secret effectively comes from A by checking the digest using A's public key, retrieved from A's certificate, then it proceeds to decrypt the secret using B's private key.

The certificates can be issued in various formats. The de facto security standard widely accepted by the industry is [X509] defined by ITU. Public and private entities relying on services (that is, trusting) provided by a common CA and accept its certificates from the PKI. Members of PKI groups can easily identify themselves to one another based on certificates provided by the CA. For this purpose the members of the PKI only need to establish secure trust relationship with one member of the PKI, the certificate authority and not with the other members. So, in short, PKI can be defined as a virtual entity combining multiple physical entities by a set of policies and rules binding public keys to the identities of key-issuing entities via the use of a certificate authority.

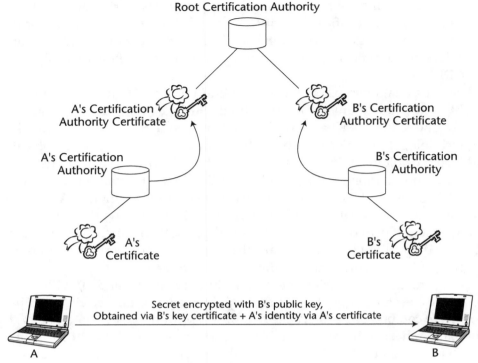

Root Certification Authority

A's Certification
Authority Certificate

B's Certification
Authority Certificate

A's Certification
Authority

B's Certification
Authority

A's
Certificate

B's
Certificate

Secret encrypted with B's public key,
Obtained via B's key certificate + A's identity via A's certificate

A

B

**Figure 2.9**   PKI based on distributed CA hierarchy.

Three main functions of PKI include:

- Certification
- Validation
- Revocation

*Certification*, or binding a key to an identity by a signature, is performed by the CA, while *validation*, or more specifically, certificate authenticity verification, can be performed by any PKI entity. The process of certification includes the generation of public key pairs, including public and private keys, generated by the user and submitted to the CA as a part of the request, or generated by the CA on a user's behalf. Validation involves checking the signature issued by CA against the CRL and the CA public key. The *revocation* of an existing certificate before its expiration date is also performed by a CA. After the certificate is revoked, the CA updates the CRL with the new information. In a typical scenario, when the user needs to obtain or validate a certificate that has been presented, it issues a request to the CA. After the requested certificate is issued or its validity verified, the appropriate information is sent by the CA to a certificate repository, which also includes the CRL.

PKI is a relatively recent networking concept defined by IETF and ITU standards and drafts. It is now being rapidly adopted by the data networking industry with virtual private networking being no exception. Authentication and key management services provided by the PKI via the use of certificates are a perfect mechanism to support stringent VPN security requirements. To use these services, VPN clients and gateways must support PKI functionalities such as key generation, certificate requests, and common CA trust relationships.

### SSL and TLS

Secure Sockets Layer (SSL) technology, originally developed by Netscape Communications Corporation, is increasingly viewed as an inexpensive and easy way to provide services similar to that provided by IP VPNs when the need to secure the IP communication between pairs of hosts is only occasional, application-dependent, or when extra level of security is required to protect the applications exchanging data. The reason is simple: SSL is usually included in commercial Internet browsers and does not require any additional software on the client side and almost no end-user involvement in the establishment of a secure connection. In recent years SSL became very popular for e-commerce applications and online banking and trading to support secure transactions requiring little user participation. For these reasons we decided to include a section describing SSL functionalities and comparing it with IPSec.

**NOTE** We must emphasize that in strict terms SSL technology has no relationship to virtual private networking. SSL can be used to achieve results similar to VPN in some cases and has therefore often been mistakenly referred to as a type of VPN or a possible near future replacement for VPN by many in the data networking community.

SSL technology is implemented in the Open Systems Interconnection (OSI) application and session layer above the transport and network layers. An SSL communication session is implemented between the SSL client, bundled with the browser's software, and the SSL server in a private network by creating a session layer point-to-point connection for applications based on TCP/IP. The establishment of an SSL session involves authentication via X.509 certificates and up to 168-bit encryption as strong as 3DES used in IPSec-based VPNs. A single SSL connection supports only one client/server application, so clients running more than one application must establish one SSL connection per application.

Recently SSL has been succeeded by Transport Layer Security (TLS), defined by [RFC2246] as the application/session layer security standard. TLS works by including TLS client software in the user application that can interface to a peer located on the server application. The TLS client and server mutually authenticate themselves via an RSA (for Rivest-Shamir-Adleman) public key encryption-based reliable protocol (using certificates) called *TLS Handshake Protocol*. During the TLS handshake phase, the peers exchange symmetric keys to be used in the communication phase, called *TLS Record Protocol*, to run a symmetric keys encryption protocol—for instance, 3DES. Just like SSL, TLS is increasingly used to support secure Web-based applications such as online banking and access to corporate services, such as email, calendaring, and secure corporate database access.

## Labeling with Multi-Protocol Label Switching

Multi-Protocol Label Switching (MPLS) is one of the best-known examples of a *labeling technology*. Labeling technologies are very similar to tunneling technologies. With tunneling, tunneled packets are delivered from a tunnel ingress point to a tunnel egress point by routers, which look up the outer header prepended to the tunneled packet. With MPLS, packets are transported from a label switched path (LSP) ingress point to a LSP egress point by looking up a label prepended to it at each hop.

Tunneling and labeling also have some differences, which Figure 2.10 highlights:

- The label has hop-by-hop significance.
- The setup of a label switched path requires setting up a label information base with appropriate information at each hop (although the first and the last hop may be statically configured with label information).

For instance, at Router B, a mapping between the incoming label, Label 1, and the outgoing label, Label 2, must exist, and a signaling protocol is required to install this information at every node. By contrast, a tunnel can be set up by Operations, Administration, and Maintenance (OA&M) at the edge of the network, and any signaling protocol does not affect intervening routers in the path traversed by packets from ingress to egress point. In addition, LSPs can be set up according to an explicit routed path, thus allowing for traffic engineering to be applied. In other words, the forwarding path can be chosen based on traffic engineering criterions rather than optimal-destination-based forwarding only. The LSP (A, B, C, and D) in Figure 2.10 may have been defined to allow a balanced usage of links from A to B.

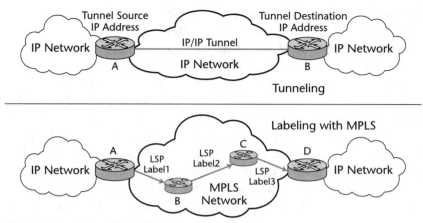

**Figure 2.10**   MPLS (labeling) versus tunneling.

Although labeling and tunneling accomplish similar objectives, they have a different—albeit sometimes overlapping—set of applications. Since each hop intervening in the path is affected by MPLS signaling and it has to keep MPLS state, then, intuitively, MPLS cannot be used with the granularity of tunneling technologies. That is, an LSP usually aggregates many more IP flows than a tunnel does. A tunnel can in fact be established between two hosts and dedicated to them only (for example, a VPN client and gateway), and carry traffic meaningful to only these two hosts. MPLS, on the other hand, is normally a router-only technology, since the granularity of an LSP is normally associated with a number of user sessions. In addition, in recognition that MPLS may present scalability issues in core networks, the standards allowed for label stacking, making it possible to aggregate a number of LSPs in the core and switch them according to a common *outer label* prepended to each individual LSP label.

Although in principle it is possible to recursively tunnel packets—that is, bundle many IP tunnels in an external IP tunnel—this is something that is not often intentionally undertaken, and it happens more by chance, when IP tunnels are carried over another IP tunnel in their end-to-end path. Notably, the label stack concept, illustrated in Figure 2.13, was not germane to the networking industry at the time of the MPLS label stacking proposal, although it was extending only to two hierarchical levels, with ATM VPI- (Virtual Path Identifier) and VCI- (Virtual Circuit Identifier) based switching. ATM allows the definition of virtual-circuit- or virtual-path-based switching. An ATM virtual path can be used to aggregate a number of ATM virtual circuits, and an ATM switch (or cross-connect, if ATM switching information is provisioned via OA&M rather than signaled) used in

VP-switching mode can simply forward ATM cells based on the VPI. In fact, one of the first specifications engineers in the IETF made when they were defining MPLS was controlling ATM switches using MPLS signaling (Label Distribution Protocol). The result of their efforts is described in [RFC3035].

MPLS was born from the fusion of multiple proposals from vendors like Cisco Systems, Ascend Communications, and IBM that defined the fundamental concepts and architecture for MPLS-based networks. The intent was to speed up forwarding by replacing IP header lookup with label lookup and augment IP with the capabilities of ATM such as QoS, traffic engineering, and different classes of service. This was the time when Ipsilon Technologies impressed the market with their *Ipswitch* product that could be used to control ATM switches to bypass the software-based routing process when an IP flow was recognized. Almost at the same time (or rather as a marketing response to it) Cisco came up with *TAG* switching, which has been, together with Ascend's IP Navigator and IBM's Aggregate Route-based IP Switching (ARIS) at the foundations of the MPLS standardization effort. The proposals on the table were a data-driven approach, where label setup is based on dynamic IP flow recognition, and a control-driven approach, where label setup is based on routing protocols, possibly augmented with some constraints and rules used to engineer traffic, and happening independently and before any packet could flow across the router. The control-driven approach was selected for standardization and shortly became the basis for the protocol development.

The MPLS architecture is based on two classes of MPLS-capable routers: label edge routers (LERs) and label switching routers (LSRs). LERs are placed at the edge of an MPLS domain and classify packets entering the domain in forwarding equivalence classes (FECs). The FEC may be based on destination IP address only, a combination of incoming (virtual) interface and destination IP address, DiffServ IP header marking, multifield classification, and other policies. The LER can be statically configured with a mapping of which label to apply to which FEC, for instance, when complex classification policies are defined locally on the LER. Alternatively, when routing protocols are driving the association of FECs to labels, the association is dynamically and automatically defined.

This property may be used to make the provisioning of multisite VPNs automatic, by applying appropriate BGP extension as described in [RFC2547]. Once packets are classified and a label is assigned to the FEC, the forwarding at the next-hop LSR of the MPLS domain will be based solely on the value of the *label*. At each intermediate LSR, a label setup protocol populates an information base that defines the association of an incoming interface and label to an FEC and an outgoing interface and label. The label setup protocol may be LDP (Label Distribution Protocol— [RFC3036]) or RSVP extended with MPLS label objects (also known as RSVP extensions for LSP tunnels, [RFC3209]). Note that RSVP and LDP both allow the setup of LSPs along engineered paths via explicit routing objects (RSVP-TE, where TE stands for Traffic Engineering) or via constraints-based routing, like route pinning or explicit routing (CR-LDP, [RFC3212]). Therefore, it is possible to maintain paths with a predictable resources level and deliver predictable service to packets entering the MPLS domain and being delivered to the domain egress LER over known and pre-engineered LSPs.

While describing these and other MPLS options, we purposely avoided a discussion on what differentiates LDP and RSVP, since this issue has always been fairly political and there are, as often happens, no huge technical advantages or disadvantages in using one or the other. There are, however, fairly significant technical differences between them in that RSVP is soft-state-based and LDP is a stateful protocol. RSVP has therefore been seen as generating more signaling traffic and being less scalable than LDP. The industry had been seen leaning toward RSVP-based label distribution for the simple reason that the dominant players in the marketplace had chosen the RSVP path (because of existing investment in this technology). However, recently, LDP seems to be the actual path the industry is following. But again, this may change over time or both approaches may survive equally successfully.

So far we have talked about the ways label information can be distributed, but we have not addressed the equally important topic of how the label information is associated to each packet forwarded between a pair of MPLS-capable nodes. There are two ways to do this. One method is associating the label to the lower-layer protocol fields (e.g., the ATM VPI/VCI fields described in [RFC3035] or the Frame Relay DLCI described in [RFC3034]). The other is inserting a header between the lower layers and the IP protocol, commonly known as the "shim header" approach, described in the MPLS label stack encoding specification [RFC3032]. Note that the shim header approach allows for stacking of labels, thus allowing aggregation of multiple LSPs in a single LSP at different level of hierarchy, which is not uncommon to the current Time Division Multiplexing (TDM) and Synchronous Digital Hierarchy (SDH) hierarchies. A label in the label stack is a 4-octets-long bits-string and it is encoded as depicted in Figure 2.11. Using multiple labels allows the bundling of LSP, as discussed earlier in the section.

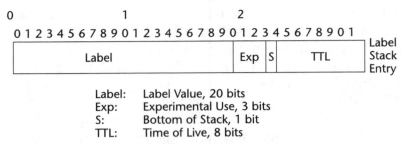

Figure 2.11   Label stack encoding header options.

Label stack encoding includes a field we need to keep in mind: the EXP field. This field is expected to be used by LSRs to classify packets belonging to the same LSP in up to eight different forwarding classes, thus enabling traffic-handling differentiation within an LSP. Notably, since up to eight classes can be defined, this mechanism allows for full compliance to the DiffServ class selector PHB (for per-hop behavior). This chapter includes a brief introduction to differentiated services in the Internet, and this PHB is discussed within it. Another proposal has been presented to use the *flow label* in the IPv6 header, which in the end does not belong to any of the classes described previously, but this method has not yet been developed further, perhaps because of the lack of the IPv6 protocol deployment.

As we end this very short overview of MPLS, we should note that the concept of label switching can be generalized in any environment. For instance, a label can be represented by a wavelength in a dense wavelength-division multiplexing (DWDM) transmission path, and having all optical routers controlled by MPLS signaling is therefore not germane. Indeed, there have been proposals to use MPLS in the MPλS form—that is, in a way that replaces the concept of labels with lambdas (λs) or wavelengths. Moreover, if we extend the domain of MPLS application to time slots, tributaries in a digital hierarchy, and other transport technologies channels or aggregate identifiers, then we have a general signaling protocol that can be used to control switching equipment of any kind (Generalized MPLS). Not limited to this, there have been recent proposals to support L2 services like Frame Relay and ATM, on top of MPLS, so that a corporation, which is used to Frame Relay-based wide area network (WAN) service for its private networking purposes could still obtain it from its current provider that is now migrating its network to be fully MPLS-based and is phasing out Frame Relay support.

# Quality of Service and VPN

The delivery of quality of service in IP networks became a very hot topic in the second half of the 1990s, because of the phenomenal explosion of IP-based service delivery in the telecommunications industry that took place after the introduction of the World Wide Web (which is the interface to the Internet most people use and has therefore now become synonymous with the Internet itself).

Differentiation of packet forwarding based on traffic classes (including voice, residential Web browsing, mission-critical, and business) or aggregates of traffic that would share common forwarding behavior within routers has proved to be the most reasonable and scalable approach to offer end-to-end QoS in the Internet. After the Integrated Services (IntServ) initiative—which defined per packet flow end-to-end services based on flow identification within each router traversed in the end-to-end path—was proved not to scale in the high-speed IP core, the IETF decided to opt for a less ambitious, but nonetheless reasonable and effective, per traffic class QoS differentiation. This approach described in [RFC2475] later became known as the Differentiated Services (DS), or DiffServ.

In this section, after a brief introduction to the DiffServ principles, we address the topic of providing QoS in VPNs based on IP tunnels and MPLS. We believe that this should be sufficient to get exposure to the kind of operational issues that a provider needs to face when running a business-grade, and possibly interdomain, VPN service.

## Per-Hop Behavior Types

DiffServ is based on packet classification at each hop—that is, at each node traversed in the end-to-end path—relying on the simple lookup of a particular field in the IP Header, called Differentiated Services Code Point (DSCP) field. After packets have been classified, a DiffServ-compliant node is expected to deliver a standard (or proprietary) PHB. Each node must be configurable to associate a PHB to a particular value of the DSCP field So far, the IETF has defined a number of standard PHBs:

- *Default* PHB—also known as Best Effort—provides for the normal packet treatment the Internet offers, defined in [RFC2474].
- *Class Selector* PHB defines up to eight configurable behaviors, which include the Default PHB. This PHB defines a backward-compatibility scheme to the precedence class mechanism [RFC791] as described in [RFC2474].

- *Assured Forwarding* PHB (AF PHB—[RFC2597]) defines four classes of service with three drop-precedence levels each, allowing for differentiation of different classes of traffic and modulation of loss and delay performance figures in each class based on the buffer space, buffer management, scheduling policy, and bandwidth assigned to each of them. The drop precedence may be changed to a higher drop precedence if, upon metering, a particular AF class traffic exceeded an agreed traffic profile. Note that an administrator may decide to mark traffic coming from different users with different drop precedence, even before the AF traffic is metered and checked against the negotiated traffic profile. For instance, it may be an administrator policy to mark employee traffic with higher drop precedence than vice presidents and higher-level personnel. The traffic conditioning agreements for AF traffic may be defined so that for each traffic class and for some drop precedence within a traffic class there is an allowed traffic level, and rules are in place to move traffic from one drop precedence to another. [RFC2697] (Single Rate Three Color Marker) and [RFC2698] (Two Rate Three Color Marker) provide examples of how these could be defined

- *Expedited Forwarding* PHB (EF PHB) guarantees a bound on the delay variation at each hop, thus allowing for a service that is suitable for applications such as circuit emulation over IP.

The delivery of predictable service in a differentiated services domain is based on well-defined rules for traffic admission to a differentiated services domain. Traffic exchanged with other domains is policed, shaped, and marked according to traffic conditioning agreements that are defined bilaterally between the domains' administrative entities. If all networks comply with the traffic conditioning agreements, and if resources within the nodes of each traversed differentiated domain are provisioned adequately, then it is possible to obtain predictable end-to-end QoS.

## QoS and Tunnels

When an IP packet is tunneled, it can traverse multiple DiffServ domains, stay within a DiffServ domain, or even transit non-DiffServ domains. These boundary conditions need to be taken into account when you are designing a tunnel-based service and QoS based on differentiated services is offered. A detailed discussion on this topic is provided in the informational [RFC2983].

In short, [RFC2983] describes two basic models:

- In the "transparent wrapper" model, the tunnel simply happens to be a transparent wrapper from a DiffServ point of view, in that the DSCP field of the inner packet is copied to the DSCP field of the outer packet header at tunnel ingress point. Then the outer IP header DSCP field value is copied to the inner IP packet header DSCP field at tunnel egress.

- In the "pipe model" the tunnel is considered a bearer service with a given QoS profile, and the DSCP field of the header of IP packets sent over it is not copied to the outer IP header upon forwarding (and, likewise, the DSCP field of the outer IP header is not copied to the inner IP header DSCP field when the packet is received at the tunnel egress). The pipe model, therefore, can be regarded as a (virtual) circuit characterized by a service profile determined by the Differentiated Services class the outer IP packet header belongs to.

To satisfy the QoS requirements of all IP flows being transported over a pipe model tunnels-based VPN, the Differentiated Services class that needs to be negotiated with the service provider must meet the most stringent QoS requirements of all IP flows carried over the pipe. This may turn out to be quite an expensive setting if the bulk of the traffic is best effort and only a small fraction of the traffic transported over the tunnel requires a high level of QoS. In this case, it may be advisable to define a bundle of pipe model tunnels-based VPN, instead of using a single tunnel-based site-to-site connectivity. You might consider this as a solution close to the transparent wrapper model. On the contrary, we want to stress that even if we had a one-to-one mapping between the user DSCP field and the outer header DSCP field at the ingress point, such mapping may change at the egress. Most importantly, the inner IP header information, such as the AF drop precedence marked at the ingress, would not be changed at the intermediate nodes. The latter property is very important, for instance, when drop precedence information needs to be preserved from the ingress to the egress of the tunnel. This ensures that the most critical traffic always has less likelihood to be dropped.

Figure 2.12 provides a synopsis of what we discussed in the preceding paragraph. The pipe model applies especially to IPSec tunnels, where for many reasons—like avoidance of traffic-analysis-based attacks—it is desirable not to copy the inner IP header DSCP field. For instance, if an attacker knew that mission-critical traffic for a particular network was marked with a given value of the DSCP field, by copying inner IP packet header DSCP field to the outer header, the attacker could discover that some packets must be collected for further analysis of mission-critical transactions. The pipe model is also useful when information such as the AF drop precedence must not be lost (note that the AF drop precedence could be lost in certain situations, such as when an intermediate domain converts all the traffic to EF). In some cases the tunnel ingress point must apply some traffic conditioning before tunneling, such as when remarking is required before tunneling packets over a pipe mode tunnel, because, for instance, the destination domain is not DiffServ-capable and only accepts packets marked according to the IP precedence model defined [RFC791].

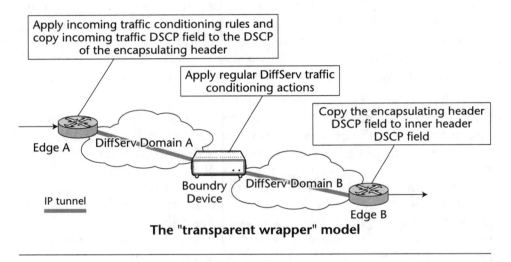

The "transparent wrapper" model

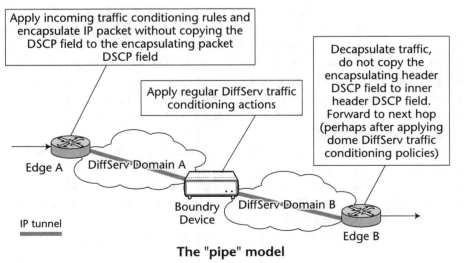

The "pipe" model

**Figure 2.12** The "transparent wrapper" and "pipe" models.

## QoS and MPLS

QoS differentiation in an MPLS-compliant network may be obtained in two ways. Using the Experimental field (EXP field) in the label stack encapsulation header allows for the differentiation of up to eight traffic classes within a single LSP. In this E-LSP approach (that is, EXP-inferred packet-scheduling class LSP), the EXP value-label value pair for incoming packets at the input interface of the LSR determines the per-hop behavior. The other approach is based on simply associating a PHB to an MPLS label value. This mode is known as L-LSP (or Label-only-inferred packet-scheduling class LSP). There is no difference in the level of QoS that can be delivered using these two different approaches. However, the ability to use the EXP field for packet classification and scheduling requires special node implementation and cannot be expected to be there in plain-vanilla MPLS QoS-capable nodes.

**NOTE** In many cases, such as ATM switches controlled via MPLS signaling, it is possible to reuse features such as the ATM Cell Loss Priority (CLP) bit to differentiate classes of traffic with different drop priorities within an MPLS LSP. In a sense, this can be regarded as a particular case of the E-LSP approach, although the terminology in this case would not be the best fit.

# Authentication, Authorization, and Accounting

The delivery of any type of services to customers by service providers normally requires three fundamental components so that the service provider can bill for service usage and deny service to undesirable customers. These components are *Authentication, Authorization, and Accounting*, commonly bundled together and identified by the acronym *AAA*:

- *Authentication* is defined as the ability or act of the service provider or attendant to request from the user of a service or resource proof of his or her identity. This service component can come in different forms and technologies, and we will detail in the next section what these may be.

- *Authorization* is the ability or act of a service provider to verify that a user whose identity is authenticated does indeed have the service access right. However, for anonymous access-based services, it may be sufficient that a user submit some credentials a trusted third party guarantees. One example may be a token-based service access paradigm, whereby an anonymous user gets access to the service simply by submitting to the service provider a token (for instance, some form of e-money or credits).

- *Accounting* is the collection of usage data that can be processed by the service provider in order to issue bills or to limit the usage of the service itself. For instance, in times of trouble or by general policy, a company may decide to put a cap on the time an employee stays connected using the corporate network free phone-based remote access. A service provider may deny authorization to a prepaid user because, based on the accounting information, he or she has exhausted the credit limit.

**NOTE** The entity providing service might not necessarily be a public network, like an ISP, but rather a corporation or an institution. In fact, for private enterprises AAA is a critical component in allowing employees access to corporate networks and protecting corporate network access from unauthorized users. In this respect, employees may be regarded as customers of the corporation, which is true also according to many modern books on management.

From the description of AAA components we have provided, it becomes clear that AAA is not only a set of three functionalities that a service provider's network usually supports but is also a set of three building blocks based on which some service logic, such as prepaid, discounting policies and network access policies, can be implemented.

## User Authentication and Authorization

The user authentication and authorization process consists of two steps:

1. Collect authentication or authorization material from the user (which we may identify as the *front-end* process).

2. Verify material (referred to as the *back-end* process). Normally, the back-end process is the service provider's part of AAA functions. The front-end mechanism is technology-dependent, and the back-end process tends to be technology-independent.

Typical examples of front-end mechanisms for collection of the user's AAA material include the PPP authentication phase Challenge Handshake Authentication Protocol (CHAP) and Password Authentication Protocol (PAP), login information and the password sent in the clear by a Telnet application client to a Telnet server at the beginning of a Telnet session, a

TLS-protected Web page where the user needs to input his or her data, and smart-card-based interfaces with Extensible Authentication Protocol (EAP) transfer of authentication information to a server that interprets the smart card information. Also, there are ways to measure biometric parameters and, for instance, identify individuals by scanning their retinas or fingerprints. In cellular networks, during the MS attach procedure (that is, when the user registers with the network its whereabouts and the network starts to manage the MS mobility) the network collects user authentication data that gets checked against data retrieved from the Home Location Register (HLR). This may come from direct user input, like in AMPS and CDMA networks, or from a chipcard known as Subscriber Identity Module (SIM) in GSM, or USIM in UMTS.

Examples of back-end process include a password file locally stored at the host device where the authentication data is collected or the query of a remote database or trusted third party for authorization and authentication decisions. In the wireless industry, for example, the approach chosen to authenticate subscribers is to download to the serving node all the necessary information to check authentication data collected from the mobile stations. Once this process is completed, the MS is then authorized to use the wireless network data services. Access to packet data networks via a wireless network may require an additional level of authentication (see Chapters 6 and 7 for some real life examples).

## Accounting Data Collection

There are two approaches for accounting data collection. One approach is storing data locally at the serving node, such as an NAS, or an application host. Drawbacks to this approach include security problems, fate sharing (if the node fails, accounting data may also be lost), and often an overly complex procedure to collect accounting data if the user is mobile and can access multiple serving nodes. The preferred approach, which is being increasingly frequently adopted in the industry, is based on sending accounting data to a gateway or server function—periodically or in an event driven fashion—and then letting the billing system access a centralized point of contact for accounting data retrieval. Of course, this centralized accounting data collection functionality requires highly reliable platforms, data integrity protection, and disaster recovery mechanisms, but all of these requirements have to be satisfied only once, while local serving nodes do not have to support such complex and specialized functionality.

Let's examine how accounting is supported in wireless systems (see Figure 2.13). In GPRS systems, accounting data is transferred to a centralized Charging Gateway Function (CGF) based on ASN-1 specified Charging Data Record (CDR) formats using the GTP'(to be read *"GTP prime"*) protocol described in [3GPP TS 32.015]. Also in GPRS, in some cases, like corporate or ISP network access, the corporation and the ISP may collect accounting data using RADIUS instead of or simultaneously with CGF, and the wireless carrier may decide to turn off GTP'-based accounting data collection, or keep it active, depending on whether flat rate or usage based bills are delivered to subscribers or on whether data is collected for trend analysis. In CDMA2000, accounting data is collected using RADIUS (extended with attributes specified in IS-835), thus sharing the same method used in data networking industry. More details on these systems is provided in Chapters 4, 6, and 7.

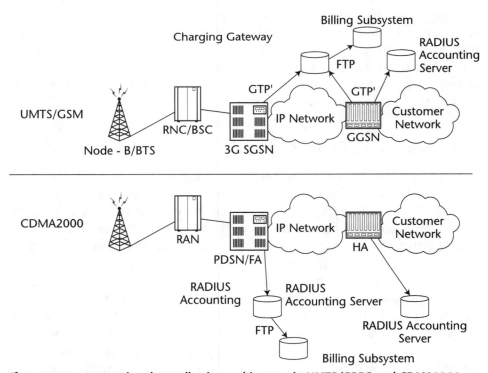

**Figure 2.13**   Accounting data collection architecture in UMTS/GPRS and CDMA2000.

# AAA and Network Access Services: RADIUS

In dial-up access as well as other network access services applications, it has been recognized that storing files at each network access server would be impractical and result in security and management headaches. Consequently, the IETF chose an approach based on a client/server protocol to perform AAA functions. The standard defines a base protocol for authentication and authorization (RADIUS—[RFC2865]) and its extension for accounting (RADIUS Accounting—[RFC2866]). The network access server collects authentication data using one of the protocols described in the following sections, and then it queries a RADIUS server by sending to it a RADIUS Access Request message to obtain a decision on whether to admit the user to the network. If the user is admitted, the RADIUS server returns a RADIUS Access Accept message; otherwise, it returns a RADIUS Access Reject. In the RADIUS Access Accept, the RADIUS server may include attributes for configuration of the network access service delivered to the user, such as the user IP address, the allowed duration of the session, and the IP address and password information necessary to tunnel PPP frames to an LNS using L2TP.

## *Authentication Methods for Network Access*

In the data networking industry, AAA has been so far largely based on the use of PPP, so most network access authentication methods have been developed to support the PPP authentication phase. The authentication phase takes place after the PPP link configuration phase has completed and before the network configuration phase starts, so that if the authentication phase is unsuccessful, no network layer configuration stage happens, the link gets turned down, and network access is denied. In the next paragraphs we consider three of the more popular methods: PAP, CHAP, and EAP.

The Password Authentication Protocol (PAP) described in [RFC1334] is a simple protocol that lets a PPP endpoint—normally a host—deliver the username and password in cleartext mode to the peer—normally an NAS—so that the peer can check their validity and allow or deny continuation of the PPP session and subsequent network access. This protocol has a weakness; an attacker may store the value of the username and password sent in the open using PAP and mount playback-based attacks on the NAS.

In response, the IETF defined a way to protect the authentication process from replay-based attacks via a three-way handshake protocol named Challenge Handshake Authentication Protocol (CHAP) described in [RFC1994]. In this protocol, a PPP endpoint can periodically challenge a peer using a random value string of bits, called a *challenge*, and the peer node must respond with a response built based on a secret shared with the *challenger endpoint*, scrambled using the value of the challenge according to an MD-5 digest function [RFC1321]. The endpoint then checks the value of the MD-5 digest contained in the response against an expected value because of its knowledge of the secret. If these results match, then the PPP session can continue; otherwise, the link is disconnected.

Flexibility in choosing the authentication protocol while the authentication phase is occurring allows authentication protocols to evolve without a software upgrade to the PPP endpoints. In fact, the selection of authentication protocols typically happens at PPP link configuration time, and the PPP link endpoints need to have hardwired implementation of all the necessary authentication protocols and the ability to negotiate them at link set-up phase. The Extensible Authentication Protocol (EAP) in [RFC2284] has been defined so that the selection of the authentication protocol can be deferred until the authentication phase and so that a back-end server can be used to implement the authentication algorithms, instead of requiring the endpoints to implement them. In fact, a PPP endpoint may be agnostic to the authentication mechanism and simply exchange EAP messages to external authentication servers implementing the authentication mechanisms. This allows you, for instance, to upgrade authentication clients on customers' hosts and servers in the operator's network without impacting the infrastructure. This may make possible a more smooth migration from password-based access to smart-card-based access and not require hard dates for the migration of all nodes. For instance, an NAS may route EAP frames to different authentication servers depending on the value of the selected authentication protocol.

The username- and password-based authentication method, such as those we just described (also known as *two-factor* authentication methods), have always been unpopular with network administrators because the maintenance of the password poses some burden on the supporting staff. In very large network deployments, frequent changes of password are counterproductive, since this normally pushes users to pick simple passwords that can be algorithmically changed (for instance, by appending two digits) at the end of a password prefix that does not change. Therefore, *three-factor* authentication methods are becoming more and more popular, since they do not require the user to manage a password. In fact, the user will be required to remember his or her identity (username) and a secret

PIN that normally is not required to be changed over time. The user then reads some digits from a secure token card and inputs them in an appropriate area in the sign-on window, or a chipcard reader appends automatically these digits that allow for the generation of a one-time password.

## AAA and Roaming: The Network Access Identifier

When a subscriber wants to use the same service as in the home network—that is, the network with which the customer-vendor relationship is in place—even when roaming to different service provider's networks (also known as *visited* of *foreign* networks), the serving visited network and the home network must be able to exchange AAA information. The way this has been addressed in the industry is to have the visited network AAA server dialogue with the home AAA server in a configuration known as RADIUS proxy. In this configuration, the visited network AAA server acts as an AAA client of the home network AAA server. This requires the visited network to be able to route the AAA messages to the right home AAA server. The IETF has laid out the method by defining a user identifier composed of the username and the domain (also known as the *realm* or *Home Network Identifier*). The format of the identifier is *user@domain*, and it is known as a Network Access Identifier [RFC2486].

When an NAI-formatted user identifier is received in an AAA message, the visited network AAA server can interpret the NAI part and route it to the home network AAA server. This model implies that there exists a trust relationship between the visited network and the home network, which in turn requires setting up security between the visited network AAA server and the home network AAA server. This does not scale well, since it has an *n2* complexity—that is, each service provider needs to set up security with all the service providers. This has been addressed by allowing third parties like clearinghouses to act as a middleman or broker. In this arrangement, the broker is the entity with which each service provider establishes a trust relationship, and this reduces the complexity to be linear. In other words, the provider needs a single security arrangement, and the complexity is offloaded to the broker, which becomes an entity specialized to handle the agreements and in the process may offer settlement services. For enhanced security, the broker may simply act as a redirector of AAA transactions, so that the visited network AAA server and the home network AAA server could be put in direct communication after the initial query directed to the broker AAA server results in redirection of the transaction to the home AAA server. This feature is not part of RADIUS capabilities and is allowed by the next-generation AAA protocol, DIAMETER. Figure 2.14 illustrates the two different models.

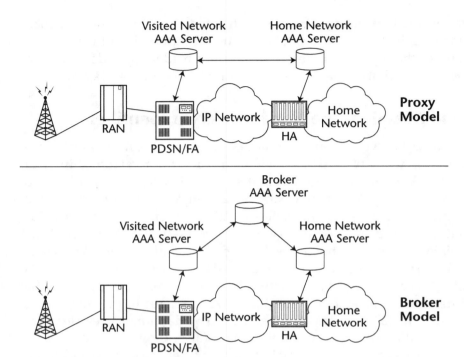

**Figure 2.14**   AAA proxy and broker architecture examples: CDMA2000.

## AAA Evolution: DIAMETER

One of the goals of the IETF AAA WG is to address deficiencies of the RADIUS protocol, both from a feature and a transport point of view. The DIAMETER protocol has been selected to solve issues related to the fact that RADIUS is a client/server-based protocol, by defining a peer-to-peer relationship between DIAMETER peer entities. This allows things like AAA server-initiated procedures. It also adds enhancements in server-to-server communication, allowing better performance in large proxy-based deployments, where server-to-server communication may lead to hundreds-of-transactions-per-second support requirements. These enhancements span the transport level, with the introduction of the Stream Control Transmission Protocol, [RFC2960]; the data model level, with hierarchical, object-oriented data models as opposed to the RADIUS flat, non-hierarchical model; and the security level, with the addition of integrity checks that would address frauds caused by malicious or unintentional alteration of AAA data by intermediate parties acting as brokers or proxies in a proxy chain. Also, DIAMETER allows for AAA brokers to act as redirect agents, thus making the communication between the visited AAA server and the home AAA server direct and mitigating the possibility of man-in-the-middle attacks.

The wireless industry is now considering DIAMETER for many other uses, such as for the interaction with a Home Subscriber Server by SIP servers in wireless networks complying with 3GPP Release 5 (R5). Work on DIAMETER is still ongoing in the IETF AAA WG at this writing, and indeed it may take some time for it to stabilize and be deployed commercially.

# Network Services

The operation of a service provider or enterprise network, including VPNs, entails some configuration and management challenges that are resolved via a set of tools, protocols, and devices. This section provides a brief overview of the operational issues and the way these are resolved via a set of tools assembled under the umbrella of network services.

## Address Management

One task that can really burden an IT administrator is assigning IP addresses to hosts. This is challenging because in assigning addresses there are constraints related to topology, and these constraints become even more challenging as the users become mobile and do not keep stationary in the same network more than a few minutes or hours. Also, the IT administrator must be mindful about conflicts between private and public addressing schemes and the use of public IP addresses in general, which recently turned into an expensive resource in short supply. So an administrator must be able to avoid conflicts and also conserve IP addresses in large deployments. Static address management does not allow for conserving IP addresses; therefore, dynamic address assignment methods must be applied. However, when dynamically assigning IP addresses, ways to check that the same IP address is not assigned to more than one user at the same time must exist.

Another problem frequently affects network administrators: Networks that do not exchange routing information, such as a private network and a public network, or a pair of private networks possibly using overlapping private addresses spaces, sometimes still need to exchange traffic. This is typical of enterprise private networks providing access to the Internet or other private networks that belong to business partners or to companies that have been recently acquired. This common problem is resolved by the use of Network Address Translation, which we address at the end of this chapter.

Of course, as the number of IP hosts in a network increases into the hundreds, the complexity of manually managing the dynamic assignment of IP addresses becomes so high that no human can reasonably be devoted to this task. This situation was at the root of the definition of the Dynamic Host Configuration Protocol (DHCP), which is commonly used in enterprise and campus networks to assign IP addresses to hosts and configure hosts with other information necessary to use the IP network services. There are other ways to dynamically assign IP address to terminals, such as the IPCP mechanism in PPP, but in this section, we concentrate on DHCP since it is the most widespread method used by landline computer networks based on shared medium and is expected to play a similar role in wireless data networks such as Wireless LANs. Another popular alternative to DHCP is PPP-based dynamic address assignment. In this method, the NAS manages local pools of IP addresses and assigns them to hosts during PPP session establishment, or the NAS dialogues with a DHCP server acting as a DHCP client on behalf of the host. It is also possible that the AAA infrastructure hands out the IP address to the NAS. Possibly, the AAA server in turn acts as a DHCP client.

### *DHCP Protocol*

The DHCP protocol, defined in [RFC2131], is a client/server protocol that allows for configuring IP hosts with an IP address and other information such as the DNS server IP address and the default gateway IP address. The protocol allows configuring an IP host without the need to have any previous knowledge of the networks where it is located. As depicted in Figure 2.15, the host's DHCP client simply sends some broadcast DHCP packets in an attempt to communicate with any DHCP server on the link where the node is attached (DHCP DISCOVER messages). Note that any router should be configured not to forward broadcast packets, but routers that are used to support DHCP—that is, routers that offer DHCP relay services— can forward DHCP queries by relaying properly modified DHCP packet onto the same subnetworks for high availability deployments. They respond via DHCP OFFER messages, each containing a server identifier.

The client receives the offers and selects one of them to use. It then responds with a broadcast DHCP REQUEST message containing the Server Identifier. Servers that are not identified by the Server Identifier abandon the transaction, while the identified server may unicast a DHCP ACK message back to the client accepting the IP address, or it sends a DHCP NACK, which notifies the client that the configuration information is out-of-date and a new configuration attempt should be started. This is useful to implement the *lease time* feature, as well as to notify clients of any inconsistent notion of configuration information with what the server assumes. A DHCP client may also release an IP address it no longer uses by issuing a DHCP RELEASE message toward the DHCP server the host has "leased" the IP address from.

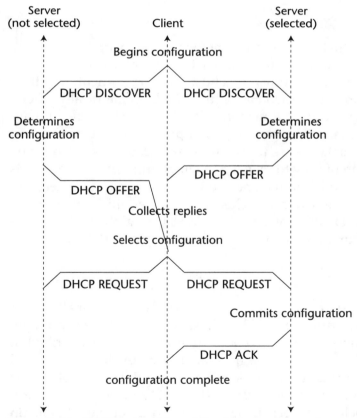

**Figure 2.15** DHCP protocol-based IP address assignment.

Also, as mentioned, DHCP may be used to configure hosts with generic information via the DHCP INFORM message the client sends to the server to get additional configuration information we do not specify here—for instance, application-level configuration information. The DHCP server responds to the INFORM message via a DHCP ACK containing the configuration information in the DHCP options requested in the DHCP INFORM message by including them. The INFORM message can be unicast to the DHCP server directly when its address is known in advance (as in the case when the host previously obtained the IP address via DHCP). Otherwise, a host can broadcast this message looking for any DHCP server to respond, which would normally happen in wireless systems when the IP address was assigned by other means or when a PPP-based host requires configuration information the Network Control Protocol (NCP) cannot deliver.

# Host Naming

One of the problems that used to affect network administrators was the need to identify an IP host via an identifier that can be easily used by a human. In fact, typing the IP address of a host is a more difficult task than writing a readable alphanumeric label, which can be semantically associated with something. Also, a mnemonic label is much more human-friendly, and since humans are the users of applications that need a host identifier as input, it makes sense to use those to identify hosts. However, this poses the problem of how to manage the mapping between the labels and the IP addresses, which in the end are what the application needs to address the IP hosts using IP packets. This drove the definition and deployment of a distributed database system that allows for a global IP hostname resolution infrastructure, called the Domain Name System (DNS).

## *Domain Name System*

The Domain Name System is a distributed database system that allows for global resolutions of hostnames to IP addresses. As illustrated in Figure 2.16, the name of a host is organized according to a syntax, described in detail in [RFC1034], that defines the hostname to be made of structured labels, made of alphanumeric strings separated by a dot. This notation is know as a *fully qualified domain name* (FQDN) and is in the form label1.label2. ... .Label(T-1).labelT. LabelT is named a Top Level Domain (TLD), and it can be a country code (e.g., us, fr, uk, it) or one of the standard

TLDs allowed for the Internet—traditionally, com, mil, gov, org, net, but more recently biz, int, nu, and others are being defined. Label (T-1) is defined as a domain, and it identifies uniquely an administratively independent namespace for which the owner of the domain has the right to define hostname-to-IP address mapping by defining an additional label value to be prepended to Label (T-1). An entity administering a domain may define the domain to be made only by a single name—for instance, bigco.com—that would map to one or more IP addresses. Alternately, the domain may be made by hundreds of hostnames obtained by adding additional dot-separated labels to the two labels identifying the domain name.

This apparently germane notation has its roots in the operation of the DNS distributed database. The DNS in fact is made of a *root zone*, which in turn is made of 12 servers that store the IP address of servers that can resolve the hostname of hosts belonging to each and every DNS domain. When a host attempts to resolve a hostname to an IP address, as depicted in Figure 2.17, it first looks up an optional internal cache. Most likely, this results in no resolution, and therefore a query is performed to a server configured for the local network. This query may fail as well—and in most cases it does, since the served names are only those associated to the IP addresses administered by the local network operator, for instance, by a certain ISP. When this happens, the DNS client on the host queries the TLD server, which points to the address of the server that can resolve the hostname. The client then can query this address and obtain from it the desired information. Normally, a single IP address is returned, but multiple IP addresses can be returned as well. The host that obtains such a list of IP addresses can use any of them to communicate with the desired server. These multiple IP addresses may be associated to IP interfaces of a single physical machine or multiple computers. It should also be noted that the DNS allows for the resolution of hostnames in at most three steps.

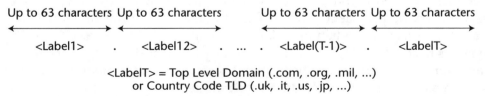

**Figure 2.16** Sample hostname structure.

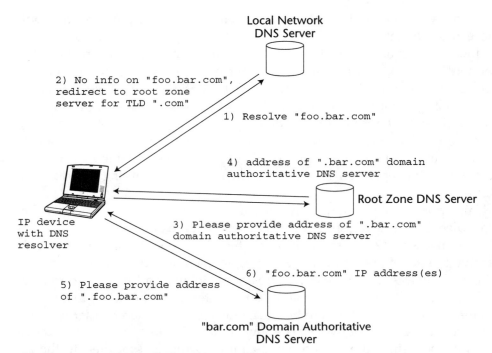

**Figure 2.17**   DNS query for hostname resolution.

It is also possible to query the DNS to find out what is the hostname associated to an IP address, without entering technical details. This functionality might be useful because a host may be able to verify the hostname of a correspondent, for instance, to check that it really belongs to a domain that has access rights to some resources. However, this feature is not often used, mainly because not every domain is configured to provide this service.

Lastly, it is worth mentioning that an organization operating a private network can define its own private DNS and its own private root zone servers serving proprietary TLDs. This is in fact the approach followed by the GSM association when they defined the .gprs TLD to be used in GPRS systems to construct FQDNs used to identify the IP address of points of termination of GTP tunnels at the GGSN. These points of termination are also known as *access points*, and hence the identifiers are known as *access point names*, or APNs. Following is the syntax of an APN:

```
APN =-<APN Network Identifier>. <APN Operator Identifier>

<APN Network Identifier> = Any valid DNS name
```

```
< APN Operator Identifier> = -<mnc<MNC>.mcc<MCC>.gprs
or -<wireless operator domain name>.gprs

<MNC> = Mobile Network code belonging to the operator

<MCC>= Mobile Country code of the Country where the operator is based
```

## Network Address Translation

Network Address Translation (NAT) is a mechanism that allows networks belonging to two independent routing and addressing realms—that is, networks that may use different or overlapping addressing schemes that are not mutually reachable based on end-to-end IP routing—to exchange traffic. For instance, the two networks may be a corporate network using a private addressing scheme and an ISP network using public addresses, or a corporation and a business partner networks both using overlapping private address space. NAT is specified by a number of IETF RFCs, such as [RFC3022] defining NAT, and [RFC2709] describing coexistence of NAT and IPSec. As defined in [RFC3022], NAT provides mapping of IP addresses from different groups transparent to the end users or their correspondents. For example, the service provider can map its private IP addresses onto publicly routable (valid) IP addresses whenever the customer needs to connect to the Internet. Technologies like NAT are needed when internal private IP addresses cannot be used outside private networks in a public addressing space either for security or compatibility reasons. The motivation behind the use of NAT is a better utilization of public IP addresses, a high-demand resource.

There are two variations of traditional NAT:

- IP Network Address Translation, also called Basic or Static NAT
- Network Address Port Translation (NAPT), also referred to as Port Address Translation (PAT)

NAPT may map multiple IP addresses to a single IP address but with different TCP port numbers. In addition, a method called NAT Protocol Translation (NAT-PT) has been proposed for translating between IPv4 and IPv6 addresses. The migration to IPv6 will be slow, and the networks will coexist for a number of years. During the transition period from IPv4 to IPv6, NAT-PT will probably be used for address translation between networks using these two protocols.

NAT works by modifying IP packet addresses so that the device doing the NAT rewrites the address headers for passed packets according to network-specific address translation rules. The packets are then routed according to the newly written header information. This restriction of remote users' access has become a major problem with VPN implementations. The reason lies in the incompatibility of NAT and IPSec, a mainstream end-to-end encryption and security technology used as a basis for many VPN services. One of the functions of IPSec is supporting end-to-end integrity of user data traffic and sometimes of the IP header (when AH is used). This means that IPSec prevents any modifications to the packet. Since the main function of NAT is to change the destination addresses of IP packets, a NAT process from the IPSec gateway and client point of view will be seen as a violation of end-to-end user data integrity and security. This can potentially disrupt end-to-end tunneling, which can have devastating consequences for voluntary VPN services. As a result, when an IPSec tunnel is established over a network implementing NAT, IPSec packet authentication will fail. Also, IKE suffers from the presence of NAT, since in IKE negotiation, parties exchange their IP address as well.

Let's consider how different modes of IPSec can coexist with NAT. In IPSec transport and tunnel modes, AH authenticates the whole IP packet, including the header. When a NAT device changes IP packet addresses, the new checksum will not be valid and the packet will be discarded at the destination. This makes IPSec AH and NAT incompatible. Contrary to AH, ESP in transport mode protects only the TCP/UDP header of the packet and not its source and destination addresses. So, in this mode the end-to-end integrity of all but TCP/UDP packets passing through NAT will not be violated. The only remaining option is ESP in tunnel mode, for which a number of solutions have been recently proposed to IETF.

One way to solve this incompatibility problem in general is based on the assumption that an NAT device is trusted by an end-to-end security association or an end-to-end tunnel will be replaced by a set of two or more tunnels concatenated or chained at NAT device. This is a simpler solution to a problem, which unfortunately somewhat defeats the purpose of end-to-end tunneling. Another approach that has been recently proposed to IETF is called NAT Traversal or NAT-T. NAT Traversal does not violate basic IPSec architecture, and it is fully compatible with standard IPSec devices. The only requirement for NAT Traversal support is the implementation of a set of capabilities, which must be supported by the devices representing tunnel endpoints and is not affecting any other devices on the data path. If during the IPSec tunnel establishment phase it is detected by the tunnel endpoints that NAT-T is supported and required, the traffic of each negotiated IPSec security association will be encapsulated in UDP packets containing the information that would help to restore the packet changed by NAT into its original form at the tunnel endpoint. The IPSec traffic between the tunnel endpoints is encapsulated in UDP using the IKE port (UDP port 500). As a result, the encapsulated packets follow the same route as IKE packets, which allows for easy firewall traversal and ensures that the modification to the packets done by NAT is similar to modification done by IKE. The original IP header length and protocol type is stored in the NAT-T header.

NAT Traversal had been proposed to IETF by SSH Communications and a coalition of Cisco Systems, F-Secure, Microsoft, and Nortel Networks. Despite still being an Internet Draft at the time of writing, this solution is already being implemented by the industry mostly because of increasing demand from the service provider community and enterprises for which NAT-T allows to establish end-to-end VPNs even if the service provider uses private addressing schemes. The control of the equipment may present a problem, as there is no way to be certain that the owners of the endpoints of the IPSec connection have any control over the equipment between the endpoints. NAT Traversal presents an easier and a much more elegant way to circumvent the problem, since it requires no changes to intermediate NAT devices—which, as stated earlier, may even be outside the control of the operator.

## Summary

In this chapter we discussed tunneling and labeling technologies. We not only explored how these could be secured via IPSec but also how security can be enforced at the application level via TLS. We illustrated the provision of QoS in IP networks and how this applies to IP tunnels and VPNs, as well as the use of AAA to enforce network access control and to offer commercial and predictable service. We concluded by addressing aspects such as address management, name resolution, and Network Address Translation (NAT), which are fundamental parts of the operation of enterprise and service provider networks, and when appropriate, we identified aspects especially relevant to wireless networks.

# Wireless Systems Overview: A Radio Interface Perspective

When we deal with wireless data, it is important to have a good understanding of both radio access and core data infrastructure aspects of this technology. The radio access technology (also known as "air interface," although we will attempt in this book to minimize the use of this term, since it is not entirely accurate) defines the technology data-capable devices must use to interface with a particular wireless system, and the associated data transmission capabilities. The core network defines the protocols used by the wireless system to interact with the mobile station for its authentication, configuration, mobility support, and session management. For this reason, we separated the background for wireless data into two chapters covering radio interface and wireless data technologies, respectively. This chapter covers the former, and the following chapter addresses the latter.

This chapter provides a brief overview of the first, second, and third generation (1G, 2G, and 3G) wireless standards for radio access. The information we present here will help you understand the capabilities of each system and the experience expectations of their subscribers. At the beginning of the chapter we attempt to clear up popular misconceptions and clarify terminology, which has been recently used loosely by many in wireless and especially data networking communities. Further in the chapter we describe the air interfaces of the more widely deployed cellular wireless systems, as well as physical properties of Wireless LANs. Wireless LANs deserve special attention for their potential in complementing or competing with 3G technologies in offering high-speed wireless data access to subscribers in low-mobility and hot-spot scenarios. This chapter provides enough background to allow you to progress confidently through the rest of the book as we cover a variety of radio access technologies. In particular, you will find this chapter useful for better understanding the material in Chapters 4, 6, 7, and 9.

## Three Wireless Generations

As previously mentioned, cellular technologies are divided into three generations: first, second, and third, abbreviated respectively as 1G, 2G, and 3G. In the past, various standard bodies and international consortia defined these terms rather vaguely and sometimes inconsistently. In response to this situation, publications and even academia are now renaming some of the technologies at the boundaries of "G" definition, adding a "decimal point" to their generation designation, such as 2.5G for example. Luckily, the discussions around this and other terminology issues in general are not entirely relevant to the main purpose of this book (and the majority of technical publications for that matter). While recognizing this problem, we do not attempt to solve it. In this book we applied current "xG.y" taxonomy as consistently as possible with as many standard definitions as we could get our hands on. Clear understanding of this classification will become important in later chapters. Table 3.1 summarizes the basic features, nomenclature, and properties of cellular systems comprising the cellular generations.

**Table 3.1**  Cellular Systems Properties

| GENERATIONS | 1G | 2G | 2.5G | 3G |
|---|---|---|---|---|
| System examples | NMT, TACS, AMPS | TDMA IS-136, GSM, CDMA IS-95, HSCSD, CDPD | GPRS, CDMA2000-1X, EDGE | CDMA2000-3X, CDMA2000-1X EV-DO UMTS, Enhanced EDGE |
| Voice/data technology | Circuit voice, circuit dial-up data | Circuit voice, circuit dial-up data | Circuit voice, circuit/packet data (Internet, IP services) | Circuit/packet voice, circuit data and high-speed packet data (multi-media, all IP option) |

**Table 3.1**  Cellular Systems Properties  *(Continued)*

| GENERATIONS | 1G | 2G | 2.5G | 3G |
|---|---|---|---|---|
| Theoretical data rate | 2.4-9.6 Kbps | 9.6 -19.2 Kbps 28.8 Kbps | 9.6 -144 Kbps; 70-473 Kbps | 144 Kbps -2 Mbps; 144 Kbps- 2 Mbps; 256 Kbps -2.4 Mbps |
| Expected average data throughput | 2.0-9.0 Kbps | 9.0-19.0 Kbps | 9.0-300 Kbps; | 60- 1000 Mbps; |
| Radio Access Technology | FDMA | TDMA, CDMA | TDMA, CDMA | TDMA, CDMA, W-CDMA, TD- SCDMA |

First-generation cellular wireless systems provide analog speech transmission based on Frequency Division Multiple Access (FDMA) with core networking based on time-division multiplexing (TDM). Examples of 1G include Advanced Mobile Phone System (AMPS), used throughout the United States, and Nordic Mobile Telephone system (NMT), which is still deployed in many Western European countries. Typically, 1G technologies were deployed within a single county or group of countries, were not standardized by international standard bodies, and were not intended for international use.

In contrast with the first generation, 2G technologies were designed for international deployment. Greater emphasis was placed on compatibility, sophisticated roaming ability, and the use of digitally encoded speech transmission over the air. This last feature is actually a requirement for a cellular system to be classified as 2G. Popular examples of 2G systems are Global System for Mobile Communications (GSM) and cdmaOne (based on the TIA [IS95] standard). 2G core networking technology may rely on circuit as well as packet data (more on this in Chapter 4).

A cellular system can be classified as a 3G system if it complies with a number of requirements set forth by the ITU:

- It must operate in one of the spectrum frequencies allocated for 3G services.

- It must provide an array of new data services to the user, including multimedia, independently from the air interface technology.

- It must support mobile data transmission at 144 Kbps for high mobility users (vehicular speed), 384 Kbps for pedestrians, and stationary data transmission at up to 2 Mbps (at least in theory).

- It must enable packet data services (that is, data services not based on a circuit connection to a data network, but on a native packet-based bearer service).

- It must adhere to a principle of independence of the core network from the radio access interface.

Some 2G systems are evolving toward at least partial compliance with these requirements. This is having an undesirable side effect: It is wreaking havoc with the "generations" terminology. For example, GSM with circuit data support is classified as a pure 2G system. When augmented with General Packet Radio Service (GPRS), it becomes compliant with many 3G requirements. This leaves it in a neither 2G nor 3G "between generations" class; as a result, GPRS-augmented GSM system is now referred to as a 2.5G system, while in actuality still belonging to a 2G class, at least from a radio transmission technology perspective.

Figure 3.1 summarizes this discussion by illustrating main cellular systems and their migration paths classified in three generations.

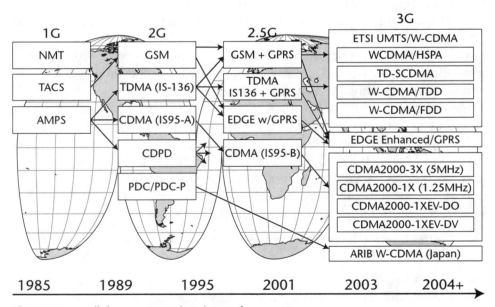

**Figure 3.1**   Cellular systems migration paths.

# 1G Cellular Systems

First-generation cellular systems have been around for a few decades now, and we expect them to remain in place for some time because of the significant infrastructure investments made by operators. All of these systems support circuit data services and may be utilized for various forms of mobile VPN, albeit not without difficulties. This section provides a high-level overview of the air interfaces utilized by most widely deployed 1G systems.

## AMPS

All 1G cellular systems rely on analog frequency modulation for speech and data transmission and in-band signaling to move control information between terminals and the rest of the network during the call. Advanced Mobile Phone System is a good example of first-generation analog technology mostly used in the United States. AMPS is based on FM radio transmission using the FDMA principle where every user is assigned their own frequency to separate user channels within the assigned spectrum (see Figure 3.2). FDMA is based on narrowband channels, each capable of supporting one phone circuit that is assigned to a particular user for the duration of the call. Frequency assignment is controlled by the system, and transmission is usually continuous in both uplink and downlink directions. The spectrum in such systems is allocated to the user for the duration of the call, whether it is being used to send voice, data, or nothing at all.

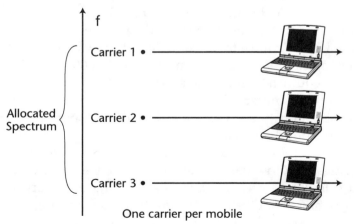

**Figure 3.2**   Frequency Division Multiple Access (FDMA) principles.

As with other 1G technologies, in AMPS a *circuit*—represented by a portion of spectrum—is allocated to the user and must remain available for this user, similar to the telephone copper pair used for voice communications. Similar to the analog wireline connection, a modem is also used for data access (see Chapter 4 for more on this). Error correction protocols used by wireless modems tend to be more robust than their landline counterparts, because of the necessity of dealing with a more challenging physical environment with inherently higher interference and signal-to-noise ratios than copper or fiber. The peak data rate for an AMPS modem call under good conditions is usually up to 14.4 Kbps, and as low as 4.8 Kbps under poor conditions. It can take anywhere up 20 seconds or more to establish an AMPS data connection.

## Nordic Mobile Telephone and Total Access Communication System

Nordic Mobile Telephone system was originally introduced in 1981 in four northern European countries—Denmark, Finland, Norway, and Sweden—in the 450-MHz frequency band, which was available at that time. Total Access Communication System (TACS) was deployed three years later in the United Kingdom and later spread to other European countries such as Italy. Both systems are based on analog FDMA radio access technology, as you would expect from a typical 1G system.

Initially, NMT was optimized for use in the sparsely populated rural environment common for Scandinavian countries. A 450-MHz frequency allowed for installation of large cells because of better propagation characteristics at higher frequencies. As business and environmental conditions changed, NMT was modified to operate in the 800-MHz range, taking into account the size and power of handsets. In contrast, TACS was designed for capacity rather than coverage. TACS systems operate in 800- and 900-MHz frequencies, which require a larger numbers of cell sites but allow for smaller and less powerful transmitters. This system proved to be very efficient and economical for countries such as the UK, with high population density and large numbers of urban areas.

Both NMT and TACS are still in use, but their customer base is eroding fast because of the widespread availability of more advanced and cost-effective GSM services. In some cases, regulators are asking operators to phase out analog (1G) systems and return the spectrum license after the subscriber numbers reach a certain low value. In these circumstances, operators, who want to retain the customer base, offer very attractively priced packages to entice analog customers to stay with the system for as long as possible.

## 2G Cellular Systems

Second-generation (2G) digital cellular systems constitute the majority of cellular communication infrastructures deployed today. 2G systems such as GSM, whose rollout started in 1987, signaled a major shift in the way mobile communications is used worldwide. In part they helped fuel the transition of a mobile phone from luxury to necessity and helped to drive subscriber costs down by more efficient utilization of air interface and volume deployment of infrastructure components and handsets.

Major geographical regions adopted different 2G systems, namely TDMA and CDMA in North America, GSM in Europe, and Personal Digital Cellular (PDC) in Japan. Figure 3.3 depicts the worldwide subscriber numbers for major 2G cellular systems. It effectively shows how the GSM system has been successful and why it is now being adopted in geographical areas other than Europe (such as North America, China, the Asia-Pacific region, and more recently, South America). CDMA, which originated in North America, has also proliferated in South America and later in the Asia-Pacific region. TDMA remains to be widely deployed in North and South America regions, but it is expected to decline mostly because of the decisions taken by few major North American carriers to convert their TDMA networks to GSM.

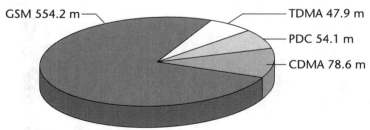

**Figure 3.3** 2G technologies worldwide market share in subscribers (2002).

## North American TDMA (IS 136)

This second-generation system, widely deployed in the United States, Canada, and South America, goes by many names, including North American TDMA, IS-136, and D-AMPS (Digital AMPS). For the sake of clarity, we will refer to it as North American TDMA, as well as simply *TDMA*, when the context makes it clear. TDMA has been used in North America since 1992 and was the first digital technology to be commercially deployed there. As its name indicates, it is based on *Time Division Multiple Access.* In TDMA the resources are shared in time, combined with frequency-division multiplexing (that is, when multiple frequencies are used). As a result, TDMA offers multiple digital channels using different time slots on a shared frequency carrier. Each mobile station is assigned both a specific frequency and a time slot during which it can communicate with the base station, as shown in Figure 3.4.

The TDMA transmitter is active during the assigned time slot and inactive during other time slots, which allows for power-saving terminal designs, among other advantages. North American TDMA supports three time slots, at 30 kHz each, further divided into three or six channels to maximize air interface utilization. A sequence of time-division multiplexed time slots in TDMA makes up frames, which are 40 ms long. The TDMA traffic channel total bit rate is 48.6 Kbps. Control overhead and number of users per channel, which is greater than one, decrease the effective throughput of a channel available for user traffic to 13 Kbps. TDMA is a dual-band technology, which means it can be deployed in 800-MHz and 1900-MHz frequency bands. In regions where both AMPS and TDMA are deployed, TDMA phones are often designed to operate in dual mode, analog and digital, in order to offer customers the ability to utilize coverage of the existing analog infrastructure.

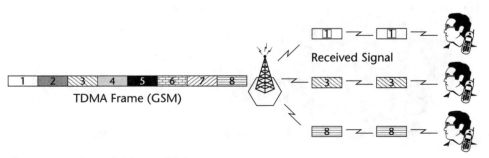

**Figure 3.4**  Time Division Multiple Access.

## Global System for Mobile Communications (GSM)

There are still some analog cellular systems in operations in Europe, but their number is declining, and some regional networks are being completely shut down or converted to Global System for Mobile Communications. The GSM cellular system initiative was initiated in 1982 by the Conference of European Posts and Telecommunications Administrations (CEPT) and is currently governed by European Telecommunications Standards Institute (ETSI), which in turn has delegated GSM specifications maintenance and evolution to 3GPP (reviewed in part in Chapter 1). The intent behind GSM introduction was to have a common approach to the creation of digital systems across European countries, to allow—among other advantages of a common standard—easy international roaming and better economies of scale by decreasing handset and infrastructure components costs through mass production. In hindsight, this was a smart political decision, which contributed to the worldwide success of European cellular infrastructure providers and equipment manufacturers.

Let's look at some details of the GSM air interface technology. The GSM standard, similarly to North American TDMA, is based on the use of two simultaneous multiplexing technologies, TDMA and FDMA. Each radio frequency (RF) channel in GSM supports eight time slots (compared to three for North American TDMA) grouped into TDMA frames, which are in turn grouped into *multiframes* consisting of 26 TDMA frames carrying traffic and control channels. Multiframes are built into *superframes* and *hyperframes*. This yields an 8-to-1 capacity increase over NMT or TACS in the same RF spectrum. The allocation of the time slots is essentially static on a short-term basis; for instance, the eighth time slot of a given RF channel is assigned to the same user each time it comes around, whether or not the user has voice or data to send.

The GSM system, emphasizing not only physical properties but also service definitions (unlike some 1G systems), supports three major types of services: bearer services, tele-services, and supplementary services. GSM *bearer* services allow for transparent or acknowledged user data transfer and define access attributes, information transfer attributes, and general attributes with specific roles. *Access* attributes define access channel properties and parameters such as bit rate; *transfer* attributes define data transfer mode (bidirectional, unidirectional), information type (speech or data), and call setup mode; *general* attributes define network-specific services such as QoS and internetworking options. *Tele-services* are what GSM subscribers actually use. They are based on the foundation provided by bearer services and govern user-to-user communications for voice or data applications. Examples of tele-services include Group 3 Fax, telephony, Short Message Service (SMS), and circuit data IP and X.25 communications. GSM *supplementary* services provide additional value-added features such as call waiting, call forwarding, call barring, and conference calling used by wireless operators to further differentiate their offerings. Further information about GSM can be obtained from a variety of sources such as [Eberspacher 2001].

### High-Speed Circuit-Switched Data

High-Speed Circuit-Switched Data (HSCSD) is an option in GSM that allows combining multiple GSM time slots (traffic channels) each capable of a 14.4-Kbps data rate. The resulting bit rate made available for a single user might reach as high as 56 Kbps, although probably at a steep price tag. In fact, owners of the mobiles capable of HSCSD support will have to pay for the combined GSM time slots being used.

Wireless carriers can achieve the migration to HSCSD by upgrading GSM Mobile Switching Center (MSC) and Base Transceiver Station (BTS) software. Wireless carriers also have to distribute handsets capable of receiving HSCSD transmission or firmware upgrades for the GSM mobiles based on Personal Computer Memory Card International Association (PCMCIA) and CompactFlash (CF) cards (such as those produced by Nokia). HSCSD can be supported within the existing GSM mobility management infrastructure, which also enables roaming and other familiar GSM services at higher data rates.

## cdmaOne

Code Division Multiple Access (CDMA) IS-95—or cdmaOne—is one of the popular 2G technologies being used in the Americas, Asia, and Eastern Europe. CDMA is based on the technique in which each subscriber is assigned a unique code, also known as pseudorandom code that is used by the system to distinguish that user from all other users transmitting simultaneously in the same frequency band. CDMA belongs to the class of systems called *spread spectrum systems*, and more specifically to the Direct Sequence Spread Spectrum (DSSS) family. Physical channels in CDMA are defined in terms of radio frequency of the carrier and a code—that is, a sequence of bits. The digital signal resulting from the encoding of voice or data, after the application of appropriate framing (or radio link layers), is digitally scrambled before it modulates the carrier frequency. This is accomplished by digitally (base 2) adding the signal to the pseudorandom code that is used to distinguish the user. The entire carrier spectrum is available to each single user, hence the name *spread spectrum*.

The receiver, which has a pseudorandom signal decoder, reproduces the original signal by demodulating the RF and adding (base 2) the same pseudorandom signal used by the transmitter, thus obtaining the original signal. CDMA is an interference-limited system, meaning that anytime a user is not transmitting and thereby not interfering with other users sharing the same spectrum, the effective bandwidth, and hence signal-to-noise ratio, available to other users will increase to some degree. CDMA properties are as follows:

- Multiple voice channels are available for each radio channel.
- To prevent interference, callers are assigned to different radio frequency channels (or, if sharing a radio channel, different pseudorandom codes).
- The same radio channel can be used in adjacent cells.
- The number of calls in a sector is "soft" limited, not hard limited.
- Bandwidth usage influences the number of simultaneous users.

To better visualize the CDMA concept, imagine a room filled with pairs of people talking to each other, each couple in their own language. They would only be able to understand their counterparts but not the rest of the conversations in the room. As the number of pairs with unique language increases, the noise level will reach its maximum, after which no conversations will be possible (not unlike in some trendy restaurants and pubs).

The CDMA cellular technology also comes with *soft handoff capability*— that is, the system specifies a receiver (*RAKE* receiver) capable of receiving up to three signals related to the same channel, because of multipath effects or to multiple sources transmitting the same signal. The system allows the mobile station to send and receive simultaneously with three base stations, which are defined as belonging to the "active set" of base stations. This allows for the avoidance of handoff Ping-Pong effects and also allows for improved performance against multipath or adverse radio conditions.

CDMA was originally deployed under the commercial name *cdmaOne* based on TIA [IS95], a mobile-to-base-station compatibility standard for wideband spread-spectrum systems. It is a direct sequence CDMA scheme in which users are differentiated by unique codes known to both transmitter and receiver. The IS-95A version of the standard allows for circuit-switched data service up to 14.4 Kbps. The next generation of IS-95, called IS-95B, requires software and hardware change in CDMA system elements and mobile stations but will support packet data at a sustained bit rate of 64 to 115 Kbps. This is achieved mostly through the use of advanced channel and code aggregation techniques and other modifications to IS-95A. In IS-95B, up to eight CDMA traffic channels can be aggregated for use by a single subscriber—not unlike HSCSD used in GSM.

# 3G Cellular Systems

In this section we provide a high-level overview of the main 3G systems radio interfaces and their properties. We address CDMA2000 and the evolution of the North American CDMA system to 3G first, and then describe UMTS and the evolution of the GSM system to 3G.

## CDMA2000

CDMA2000, often called CDMA 3G, is a next generation of original CDMA IS-95A and IS-95B cellular systems. It has been already deployed in America, as well as some regions in Asia and Eastern Europe. The CDMA2000 Radio Transmission Technology (RTT) includes enhancements that effectively double CDMA IS-95 spectral efficiency, as well as the number of simultaneous voice calls the system can handle. CDMA2000 uses transmission chip rates, which are multiple of 1.2288 Mcps, which means that CDMA2000-based cellular systems are backward-compatible with the large installed base of CDMA IS-95 systems and CDMA2000 mobile terminals can be made compatible with legacy systems. CDMA 2000 deployment can support multiple carrier bandwidths and frequency band plans including PCS and IMT2000. The CDMA2000 RTT support for 1.25-MHz deployment capability is especially important in North America, where it enables CDMA carriers to reuse their existing spectrum to offer CDMA2000 service by only upgrading the equipment without the necessity to buy new spectrum. This version of CDMA2000 had been named CDMA2000 phase one, or CDMA2000-1x RTT. The other version of CDMA2000 technology, called CDMA2000-3x, representing the logical evolution for CDMA2000-1x, employs 5 MHz of bandwidth.[1]

CDMA2000-1x will provide subscribers the ability to transfer and receive packets at raw data rates of 153.6 Kbps or an effective data rate of up to 144 Kbps. CDMA2000-1x, just like its predecessor, provides support for both voice and data. The CDMA2000-1x physical layer incorporates a number of major enhancements that provide for higher data rates and better spectral efficiency than the second-generation CDMA systems. A

---

[1] Hence the name: 1x for 1.25 MHz and 3x for 5 MHz, which is roughly three times that.

burst-mode capability is defined to allow better interference management and capacity utilization. An *active* high-speed packet data mobile always has a traffic channel using a *fundamental code*. This channel is called the *fundamental channel (FCH)*. An active High Speed Packet Data (HSPD) call with the need for higher bandwidth either in forward or reverse direction could be allocated an additional channel for the duration of a data *burst*, which is in the order of seconds. The additional channel during this state is called the *supplemental channel* (SCH), which allows a wide range of data rates; raw data rates of 9.6 Kbps to 307.2 Kbps are supported over each SCH. Though the numerology in the proposed CDMA2000-1x standard supports raw SCH data rates as high as 307.2 Kbps, it is anticipated that the maximum effective data rate that can be supported over a wide area of coverage—with a single 1.25-MHz carrier—will be around 150 Kbps. Usually one SCH is assigned per data service. An SCH with a data rate of 19.2 Kbps or higher is equivalent to multiple voice calls from the consideration of air interface capacity.

## CDMA2000-1xEV

To meet the wireless data market requirement for a high-speed, high-capacity solution, the CDMA standard community has developed a data-optimized version of CDMA2000 called CDMA2000-1xEV-DO (where DO stands for *Data Only*). CDMA2000-1xEV-DO supports downlink (to the MS) peak data rates of up to 2457.6 Kbps, and 153.6 Kbps data rate on the uplink (from the MS). Note that while CDMA2000 systems had been designed to provide equal upstream and downstream capacity, the 1xEV-DO version of CDMA2000 had been designed to accommodate asymmetric needs of high-speed data users, which require more bandwidth in the downlink direction for downloading information from the Internet and receiving streaming media traffic. Packet data is the only type of traffic currently supported by this technology. Voice over IP support will be added later as the standards mature.

1xEV-DO will provide users with "always-on" packet data connectivity, not unlike current DSL and cable modem offerings. A 1xEV-DO network can be deployed as complementary to a regular CDMA2000-1x network, since most of the hardware components as well as spectrum can be shared between the two. When planning a combined system, you must allocate a separate dedicated 1.25-MHz frequency carrier for 1xEV traffic. While two systems can coexist in a carrier's network, they will require different handsets or two different functionalities supported in one handset in future deployments.

### CDMA2000-3x

CDMA2000-3x (or CDMA 3G-3xRTT) uses 5 MHz of bandwidth, and it is therefore classified together with UMTS in the Wideband CDMA (W-CDMA) family of radio transmission technologies. It delivers peak bit rates of up to 144 Kbps for mobile applications and as much as 2 Mbps for stationary applications. CDMA2000-3x will also introduce higher bit rates for data transmission, more sophisticated QoS and policy mechanisms, and advanced multimedia capabilities. It will rely on the ATM-based data link layer between the base stations and MSCs to accommodate the higher speeds and advanced call model. Table 3.2 shows a comparison between the CDMA technologies, including the UMTS W-CDMA technology, which is described in the following section.

## Universal Mobile Telecommunications System

The Universal Mobile Telecommunications System (UMTS) is a third-generation cellular wireless system compliant with all third-generation system definitions. UMTS was designed to be an evolution of the GSM/GPRS system. As a result, UMTS inherits most of the GSM/GPRS core network architecture, and UMTS and GSM roaming and intersystem handover with dual-mode handsets capable of supporting both UMTS and GSM are guaranteed. The UMTS system is defined by 3GPP in three releases: R99, R4, and R5 (see Chapter 4). The UMTS system shares the core network with the GSM system, and from R99 the specifications for the GSM and UMTS core are the same, as mentioned in Chapter 1. Release 99, however, introduces an entirely new radio interface standard based on W-CDMA technology, as well as a new method of interfacing the core to the radio access network (RAN). This chapter provides a brief overview of the UMTS radio interface, while Chapter 4 offers a detailed overview of its core network.

**Table 3.2**   CDMA2000 Technology Comparison

| | CALLS PER RADIO CHANNEL | RADIO CHANNEL BANDWIDTH | THEORETICAL PEAK DATA RATES |
|---|---|---|---|
| cdmaOne IS-95 (A/B) | 13-28 | 1.25 MHz | 9.6/14.4/19.2 Kbps |
| CDMA2000-1X 1XEV-DO | 140-200 | 1.25 MHz | 144-153.6 Kbps<br><br>144 Kbps - 2.4-Mbps peak |

**Table 3.2**   CDMA2000 Technology Comparison   *(Continued)*

|  | CALLS PER RADIO CHANNEL | RADIO CHANNEL BANDWIDTH | THEORETICAL PEAK DATA RATES |
|---|---|---|---|
| CDMA2000-3X | 180-300 | 5 MHz | 385 Kbps 2.4-Mbps peak |
| W-CDMA | 140-200 | 5 MHz | 144 Kbps—pedestrian 384 Kbps—vehicular 2.4 Mbps—stationary |

### UMTS Standardization

UMTS has been specified by International Telecommunications Union (ITU) in the context of the International Mobile Telecommunications System (IMT-2000) initiative. The 3G Partnership Project (3GPP) was formed in 1998 to coordinate a global effort by a large variety of regional standardization bodies to define the specifications for the UMTS system. The bodies involved in the specification included Association of Radio Industries and Businesses (ARIB) from Japan, Telecommunication Technology Committee (TTC) from Asia, T1P1 from TIA (North America), and European Telecommunications Standards Institute (ETSI) from Europe. ETSI also gave 3GPP the task of maintaining and evolving GSM specifications. This was the logical result derived from the use of the GSM core network as the basis for the development of the UMTS core network specifications and maintaining full roaming and interoperability between GSM and UMTS systems. Network operators and manufacturers also came together to form the Operators Harmonization Group (OHG), which encourages mutual interoperability and roaming between the CDMA2000 and wideband CDMA (W-CDMA) standards—with the ultimate intent, we can predict, to eventually converge to a new globally accepted international standard some time in the future.

UMTS supports the concept of *Virtual Home Environment* (VHE), which allows subscribers to access the same services from anyplace in the world where UMTS is supported. Subscribers, using the systems supporting VHE, experience the same look and feel of the user interface and will be able to use the same services as in the home network while roaming in any other Public Land Mobile Network (PLMN). The details of the services will largely depend on the capabilities of the visited network and service agreements set up between the various operators. Like its predecessor, the UMTS system offers circuit- and packet-based services. The core networks sections that support these services are called UMTS CS domain and UMTS PS domain. Finally, UMTS offers a core network that is independent from the RAN. This is accomplished by hiding cell level mobility aspects that used to affect the core in GSM inside the RAN. In UMTS, for instance, the core is not aware of the location of the user on a per-cell basis.

### UMTS Radio Interface

The UMTS network is logically divided into a UMTS Terrestrial Radio Access Network (UTRAN) and a core network (CN), connected via an open interface (the Iu interface) as shown in Figure 3.5, which depicts a simplified 3GPP reference model. When used to interface with the CS domain, the Iu interface is called Iu-CS, and when it is used to interface with the PS domain, it is called Iu-PS.

**NOTE** The term UTRAN is used instead of simply URAN because the UMTS system, complying with the IMT-2000 recommendations, was supposed to allow the support of user mobility not only through *terrestrial* radio equipment (antennas that nowadays we frequently spot on highways and building rooftops) but also through satellite equipment. The terminology reflects this by adding "Terrestrial" to further qualify which access network the UMTS core the specifications refer to. We suspect, a bit cynically, that this other aspect of UMTS might have been safely skipped—given perhaps the less-than-stellar results from the attempts to deploy satellite-based systems.

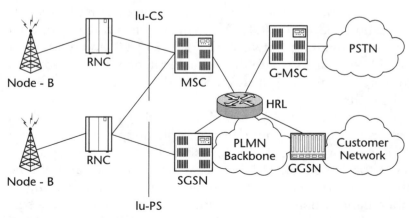

**Figure 3.5**   3GPP UMTS architecture.

From a data service perspective, the GPRS and UMTS PS domains are similar—both allow wireless service providers to offer bidirectional packet data service. The GSM and UMTS CS domains allow for both circuit data support and speech services. The UMTS system allows for higher data rates and multimedia support over the CS bearers, that the GSM did not.

On the high level, UTRAN consists of the Radio Network Controller (RNC) and Node B. RNC is functionally similar to the Base Station Controller (BSC) in GSM networks, although with greater intelligence and a number of new responsibilities because of the fundamental principle of independence of the core from the RAN, which requires the RAN to hide many user mobility support aspects that were originally handled by the core. The RNC manages the radio resource functions, such as call establishment and release, power control, and soft handover. An RNC also triggers the relocation of an MS to another RNC, when it is optimal for an MS to be managed by a new "serving" RNC. This may also involve changing the core network node the MS is homed to, or the CN node may stay unchanged. Two radio interfaces are specified for UMTS: one is a Frequency Division Duplex (FDD) operation using FDD/W-CDMA radio technology, and one is based on a Time Division Duplex (TDD) operation using TDD/W-CDMA radio technology.

UMTS licenses allocate certain frequency bands to operators. One type of allocation is known as *paired spectrum*. Paired spectrum uses two separate frequency bands, one for uplink and one for downlink traffic. This allocation method is used by the FDD operation. The other type of allocation, used by TDD, is called *unpaired spectrum*. Here mobile stations send and receive information on the same frequency. Because very fine timing control is required for sending and receiving packets using TDD, this is more suitable for shorter-distance applications, such as the indoor "picocells" that will be installed in places with high user concentration, such as airports, offices, and trade shows.

UMTS will support higher-speed data transfer across the air interface than its parent, the GSM system. It defines three mobility categories with corresponding bit rates:

- *High mobility*, such as communication from an automobile or a train with up to a 144-Kbps data rate
- *Low mobility*, such as communication while walking, with up to a 384-Kbps data rate
- *Indoor mobility*, which is not really a true mobility and constitutes communication while stationary, with up to a 2-Mbps data rate

The actual data rates will depend on the capacity of the air interface and the environmental conditions at any given moment. A larger number of high data rate users at the air interface will substantially reduce the overall throughput they will experience from the peak rates we just listed.

## Enchased Data Rates for Global Evolution

Enchased Data Rates for Global Evolution[2] (EDGE) is still considered by some a 2G technology; however, given the latest standardization directions we are about to discuss, EDGE, at least in its next-generation versions, may very well deserve a 3G status. This prompted us to include it under the 3G section of this chapter. EDGE was conceived in 1997 as an evolution path for GSM only, but was later adopted for North American TDMA. The adaptation of EDGE as a possible air interface technology evolution for North American TDMA took place in 1998 when the main TDMA technology industry forum, the Universal Wireless Communications Consortium (UWCC), adopted it as a basis for the 136 high-speed (136-HS) standard to provide 384-Kbps services, which helped to satisfy many of the 3G system requirements.

While EDGE reuses GSM and TDMA *spectrum* and *time slot structure*, it is not restricted to GSM *channel structure* and is based on modern, highly efficient modulation schemes. The EDGE radio technology relies on a fundamental *link quality control*, a concept that allows for adaptation of data protection to channel quality, thus achieving the optimal bit rate for a variety of radio environments. EDGE modulation schemes allow for the total bit rates from 22.8 Kbps up to 69.2 Kbps per time slot, respectively. A maximum of eight combined channels in the EDGE system with an estimated effective bit rate of 48 Kbps per channel will yield as much as 384 Kbps of total throughput (theoretically up to 473.6 Kbps), which is in line with 3G bit rate requirements for wireless packet data. Basically, EDGE only introduces new modulation techniques, new channel coding, adjustments to radio link protocols, and other enhancements to the existing GSM and TDMA systems without effecting their core networks.

**NOTE** EDGE technology provides the evolution path for the GSM and TDMA air interface only, while the packet core infrastructure remains based on GPRS. That's why we have not included EDGE in the following chapters, which discusses core infrastructure and wireless data delivery mechanisms.

### EDGE Classification

A few different flavors of EDGE exist, as shown in Figure 3.6. We will briefly describe each one.

[2] Originally the "G" in EDGE referred to GSM but now is increasingly replaced with Global to be more inclusive and accurate (EDGE is also an evolution path for TDMA). All three terms are still being used, however, depending on who is presenting them.

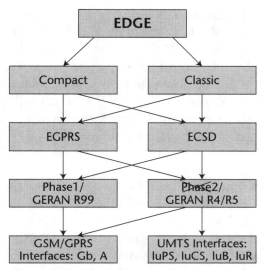

**Figure 3.6**   Edge taxonomy.

First are the EDGE Compact and EDGE Classic versions. EDGE Compact designed for TDMA IS-136 uses 200-kHz control-channel structure. EDGE Compact allows for data rates of up to 7/8's that of ETSI-defined EDGE classic. It uses synchronized base stations to maintain a minimal spectrum deployment of 1 MHz in a 1 frequency-reuse pattern. EDGE Classic employs the traditional GSM 200-kHz control-channel structure with a 4/12 frequency-reuse pattern on the first frequency.

Second, there is a distinction between EDGE systems deployed for packet-switched data enhancement called Enhanced GPRS (EGPRS) and circuit-switched data enhancement called Enhanced CSD or Enhanced HSCSD (ECSD or E-HSCSD).

Finally, there are two EDGE phases. Phase1, also referred to as GERAN R99, and Phase2, now superceded by the name GERAN R4/R5. EDGE Phase1 defines new air interface capable of up to 384-Kbps packet data throughput and will require smaller cell sizes compared to those of GSM. It also defines single- and multislot packet-switched and circuit-switched services. EDGE Phase1 standardization has recently been completed within 3GPP Release 99. EDGE Phase2, which is still under standardization, defines the alignment with UMTS (see next section) under the concept known as GSM/EDGE Radio Access Network (GERAN). In fact, EDGE Phase2, or GERAN R4/R5, will extend all the services and throughput of the Phase1 over UMTS core network. To achieve that, the GERAN R4 and R5 standards framework is mostly centered around the advanced services definition similar to UMTS services architecture.

The eventual objectives of GERAN specification include providing classes of data services similar to UMTS and the ability to interface and handover with UMTS CN over an Iu interface. While GERAN R4 introduces only minor enhancements to the current EDGE standard, mostly around the air interface, and retained the interface (Gb) for connection to GPRS, major enhancement packages are expected to come with GERAN R5, which will introduce the UMTS QoS framework, support for the Iu interface and protocol architecture, and other enhancements related to the radio interface.

### The Future of EDGE

Designed as a GSM and TDMA enhancement, EDGE will efficiently coexist with the current cellular infrastructures based on time-division multiplexing and potentially can provide a migration path alternative to UMTS. EDGE might be attractive to those wireless operators that were unable to obtain expensive 3G licenses—for instance, in some Western European countries. These operators expect the combination of EDGE and GSM to provide a path to the offering of global roaming and 3G services, while not requiring the acquisition of new spectrum necessary to run a 3G system based on W-CDMA, somewhat similarly to the 3G-1X approach used by traditional CDMA carriers. The migration to EDGE in North American markets is even more likely, because the UMTS spectrum is not available in this part of the world.[3]

# Wireless LAN

If this book had been written at the end of 2000, wireless LAN (WLAN) would most likely not appear here. After all, WLAN has only superficial similarity to cellular systems, does not truly support data mobility unless tunneling schemes like Mobile IP are applied to it, and was originally designed for in-building or campus rather than wide-area deployment. But since this time, WLAN has arguably become the fastest growing wireless access technology. We believe WLAN—combined with 2.5G and 3G systems—can potentially bring wireless data communication to a new level unachievable by 3G alone. For instance, a number of hotels and café chains in many parts of the world, such as Western Europe and North America (for example, Starbucks in the United States), already offer their guests the ability to use WLAN services. These businesses often partner

[3] A different version of UMTS radio interface operating in a frequency different than the European system would be the only way to deploy UMTS in North America unless the spectrum becomes available.

with a cellular operator or wireline ISP, which manages the service for them, or asks the operator for permission to install their system at the premises based on a lease agreement or a revenue-sharing approach.

## WLAN Technology

WLAN is a version of the existing wireline LAN standards (802.3 or Ethernet are two popular examples) defined by IEEE named 802.11b, after the number of engineers assigned to design it. WLAN relies on the RF technology and transmits data over the air, rather than via the wired connection required by other LAN technologies. In this section we briefly review the properties of the air interface used by WLAN. Networking and security are discussed in detail in Chapter 9.

Successful WLAN connection requires a client device, a wireless Network Interface Card (NIC), installed in the user's mobile device (such as laptop or PDA), and an access point (AP), the device that terminates the RF air interface and performs routing functions. A typical WLAN topology is shown in Figure 3.7.

WLAN ranges can reach as high as 500 meters outdoors with no obstacles present. It is based on use of unlicensed Industrial, Scientific, and Medical (ISM) band, 2.4 to 2.5 MHz, which brings clear cost advantages to commercial WLAN deployments. On the downside a 2.4-MHz frequency is now also used by cordless telephones and Bluetooth devices and can potentially interfere with microwave ranges emission and vice versa.

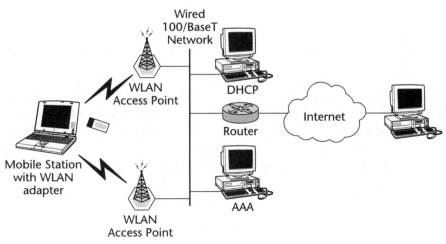

**Figure 3.7**   Typical WLAN topology.

The networks defined by 802.11 standards are capable of operating by using two types of transmission: Frequency Hopping Spread Spectrum (FHSS) and Direct Sequence Spread Spectrum, already mentioned in the CDMA2000 section. FHSS and DSSS radio interface technologies are mutually exclusive. FHSS is simpler to implement but is limited to 2 Mbps by FCC regulations (in the United States, where it was developed). For unlicensed band, DSSS is more complex but offers higher tolerance to interference and the bit rates as high as 11 Mbps, which is the highest-possible bit rate for the 802.11b standard. As such the DSSS version is largely the most popular and successful, and it is fast becoming the de facto standard.

One of the major limitations to wider deployment of WLANs for public and corporate use has been the lack of security or a standard way to secure the access and data confidentiality. The Wireline Equivalent Protocol (WEP) standard developed by IEEE tried to address this, but examples of failure made it unsuitable for the deployment in networks with strict security requirements. The interoperability between different WLAN standards is now being addressed by Wireless Ethernet Compatibility Alliance (WECA) initiative, which is now promoting not only interoperability among vendors (via the definition of the Wi-Fi trademark and logo) but is also closely monitoring security extension to WEP that IEEE is currently defining in the 802.11i committee. Typically, public WLAN deployments are based on the users authenticating at a portal, possibly after TCP redirection occurring as the user attempts to access any Web page. (An example is the *captive portal* approach, common to many broadband and now also wireless access systems; see Chapter 5 for more details.) After authentication, the point of attachment of the user to the Internet will allow the user access until further authentication is requested or an account needs to be *topped up*.

## Summary

In this chapter we introduced cellular systems radio interface fundamentals. We expect this to be a basis providing an understanding of the systems capabilities, and terminology as we will use the numbers, references, and acronyms in the ensuing chapters. In the next chapter we will address cellular systems core network aspects to complement the discussion in this chapter and prepare for the second part of the book.

# Wireless Systems Overview: Data Services Perspective

Late first-generation (1G) and second-generation (2G) technologies—such as Code Division Multiple Access (CDMA) and Global System for Mobile Communications (GSM) cellular systems—have been supporting circuit-switched data services for a few years now, providing users with low-speed connectivity. While reasonably popular and generally ubiquitous, circuit-switched wireless data services have some serious drawbacks, including poor utilization of radio resources and limited throughput. Packet technologies, introduced within CDMA2000, GPRS, and UMTS systems, were designed to address the majority of these limitations.

In this chapter we provide an overview of current and upcoming wireless packet and circuit data systems to provide a solid foundation for the Mobile VPN discussion in the second part of the book. We begin with the analysis of circuit versus packet wireless data, which we follow with an overview of data services delivery mechanisms within 1G and 2G systems. The second half of the chapter is entirely devoted to a detailed discussion of packet data services within the 2.5G and 3G systems framework. Note that we have deliberately chosen not to describe aspects of the systems that we perceive as not central for the purpose of this book or not likely to be a part of any real-world deployment.

# Circuit versus Packet

With traditional *circuit-switched* wireless data, dedicated circuits are allocated to subscribers whether or not they are being used. In theory, this provides higher effective bandwidth to users by dedicating a full channel to each user. The data service is provided via a wireless dial-up model, similar to wireline dial-up remote access. The user dials a phone number associated with the network access server (NAS) used for a specific wireless data service. Once the physical connection—that is, the circuit—is established between the mobile station and the NAS, PPP is used to provide end-to-end link layer service. Terminating the user PPP session can be easily accomplished using simple dial-up techniques based on off-the-shelf modem banks or remote access servers (RAS), comprising an Interworking Function (IWF) functionality, like the 3Com (now its spin-off, CommWorks Corporation) Total Control RAS chassis, with some software upgrades making it suitable for wireless environment. The IWF is normally required to terminate the wireless access protocols (Radio Link Protocol, or RLP) and interwork with the Public Switched Service Telephone Network (PSTN) when needed. In some cases the IWF also can relay PPP to a private network using L2TP (details on this and other options are given in Chapters 6 and 7).

In contrast, wireless *packet-switched* data technologies are based on a wireless access network support for statistical multiplexing of user sessions over the radio interface. Packet data network resources are only consumed during the transfer of the data and unused during idle periods, resulting in a more efficient system in which any source of traffic can use resources not used by others. Known as *statistical multiplexing*, this is an important property of all packet data systems. Statistical multiplexing makes packet data systems more efficient than circuit-based systems, which guarantee each user a separate, dedicated channel, not fully utilized with bursty data transmissions patterns. However, it also means that users of shared media networks must contend for the available bandwidth, sometimes resulting in congestion, delays, and lower effective throughput per user.

Access contention for shared resources is a typical problem not only for cellular packet environments but also for Wireless LAN. In cellular systems supporting packet mode access, to make efficient use of resources, the radio access bearers are only temporarily allocated to a specific user. After a period of inactivity, the mobile station enters an idle (for instance, in GPRS) or dormant (CDMA2000) mode of operation. This mode allows the mobile station to be constantly reachable by signaling and data sent to its network layer address using location update procedures and paging, while no dedicated resources are active to allow the MS to send and receive data. When data needs to be received, the MS is paged, "wakes up," and issues a request to set up the radio bearer that would allow data reception. The MS issues the same request when it needs to send data and when no radio bearer is already set up.

Packet data user *mobility* support is conceptually similar across different wireless data systems. It is based on various tunneling mechanisms such as Mobile IP (adopted by CDMA2000) and GTP (adopted by GSM and UMTS), both of which are analyzed later in the chapter. This common packet data tunneling model is shown in Figure 4.1. Tunnels depicted in this figure by thick dashed (for former, discontinued) and solid lines (for current, active) are dynamically established between the mobile station's temporary point of attachment to the wireless network and a tunnel "anchor point" or home network that also acts as a gateway to the mobile data network from which the user is receiving access service. As mobile stations dynamically change location within the network—for example, traveling through certain geographical area from Mobile Switching Center (MSC) to MSC or being at the MSC boundary—tunnels are dynamically established between the MS home network and the visited wireless access network.

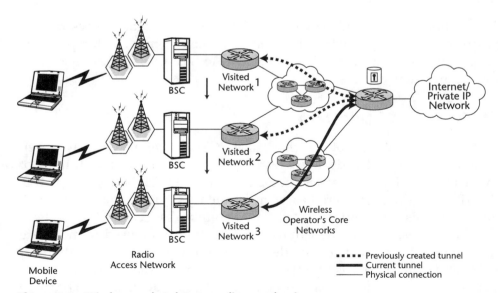

**Figure 4.1** Wireless packet data tunneling mechanism.

The majority of early data users familiar with using wireline dial-up remote access services had little or no trouble adapting to the wireless circuit-switched data services. This helped in driving circuit data acceptance rate fairly high, especially considering its low bandwidth and other limitations, which limited the willingness to frequently use the service and to stay online for a long time. All the familiar wireline dial-up access features were also present in wireless circuit data: familiar login-password sequence, ability to access corporate network by simply dialing specific telephone numbers, identical configuration procedures on user client devices such as laptop computers.

A relatively similar pattern of service adoption for wireless packet data technologies such as GPRS has emerged, although we have recorded some longer-than-expected service take-up periods mostly caused by the lack of volume terminal production and pricing experiments by wireless carriers. Wireless packet data technologies do not require dialing a number to reach a specific NAS in a corporate or ISP network. Instead, users enjoy the simplicity of *always on* or *on-demand* connectivity to the Internet or their corporate network. However, in most cases, this does require predefined relationships between the corporation and wireless carrier, as opposed to circuit data networking.

# Data Services in 1G, 2G, and 3G Systems

This section gives a detailed overview of how wireless data services are provided in the most popular cellular environments. Similar to the previous chapter, we have broken up the information according to first-, second-, and third-generation systems.

## 1G Systems Circuit Data

Cellular and PCS circuit-switched data (CSD) technologies have been around for over a decade. VPN techniques can be utilized with most of these systems. 1G wireless technologies such as TACS, NMT, and AMPS, described in Chapter 3, are based on frequency-division multiplexing. A voice signal is transmitted directly on a chunk of spectrum that is dedicated to the user. This same chunk of spectrum can be used to modulate data signals using a wireless modem, similar to what happens with the 3-kHz band of telephone copper pair used for modem dial-up data connection. Error correction protocols used by wireless modems tend to be more robust than their landline counterparts because of the necessity to deal with a more challenging physical environment with inherently higher interference and signal-to-noise ratios than copper or fiber cable.

The data rate for an AMPS modem call under good conditions is as high as 14.4 Kbps and under poor conditions is as low as 4.8 Kbps. There may be a delay of 20 or more seconds to establish a connection. As a result, the user experience is not entirely satisfactory, and combined with the poor price-performance ratio, this has made the CSD services over first-generation cellular systems an insignificant source of revenue for carriers. This partially explains the certain stagnation of data services during the 1990s relative to voice in the middle of a heyday of 1G cellular systems.

## Circuit-Switched Data in 2G and 3G Systems

Circuit-switched data has not vanished with the first-generation systems. Instead, it made a logical transition and is now available as an option in both 2G and even upcoming 3G systems. Second- and third-generation systems are based on digital transmission of speech over dedicated radio resources (either TDMA time slots or a code in CDMA, see Chapter 3). These systems are based on the use of digital channels, and therefore modulation and demodulation do not need to occur in order to transmit data from the mobile stations to the core network. These systems have been augmented with data circuit-switched services capabilities and, as a result, may offer transparent data service or "reliable" service by using Automatic Repeat Request (ARQ) mechanisms, which involve retransmission of lost chunks of data transmitted over the digital channel. Also, these channels need to be terminated at some place in the wireless network either to provide direct access to a data network or to interwork with ISDN or PSTN lines, in order to provide end-to-end connectivity to an IWF in the wireless carrier's core network or a router in a network owned by another party, such as a corporation or an ISP. In the following sections, we explore the capabilities offered by GSM, UMTS, CDMA (IS95), CDMA2000, and TDMA (ANSI-136).

### CDMA and TDMA Circuit-Switched Data

Circuit-switched data service in both North American TDMA and CDMA IS-95 systems is provided similarly to other 2G cellular systems. Shown in Figure 4.2, this service requires a fixed circuit—in this case, a CDMA modem dial-up connection—to be established between the mobile and the call destination. The system must carry the data in a digital form to an IWF, which can then generate the modem tones for communication through the PSTN.

**NOTE** In fact, the users who subscribed to a circuit data service offered by a wireless carrier can simply use their cell phones as terminals extending physical layer to connect to external modems in IWF, often supported within the MSC. An IWF is effectively splitting the mobile data call into land and mobile segments. The 2G mobile phone, and in fact the radio network (including the cell tower, base station, and MSC), acts as a conduit for passing commands to the PSTN-connected modems in the IWF. In a sense, this system provides a *virtual serial cable* (similar to that which would connect a computer and an external modem on the desktop in a landline remote access scenario) that extends from the mobile computer to the modem in the IWF.

To this effect, a data call originates at the IWF in service provider networks, which dials out to the mobile's destination number. IWF converts CDMA or TDMA -encoded data into normal analog modem format and passes the call over the PSTN. CDMA airlink supports two data rates: 9.6 Kbps and 14.4 Kbps. Additional compression techniques often allow users to achieve data rates up to 19.2 Kbps. The call sequence for a typical CDMA circuit data connection includes the following steps, which are illustrated in Figure 4.2:

1. The circuit connection is established by the MS.
2. The data call is directed by the MSC to the IWF.
3. The outbound call is placed by the IWF modem over the internal IWF packet bus to a TDM switch.
4. The TDM switch connects the call through the PSTN.
5. The call is terminated by the modem at the call destination.

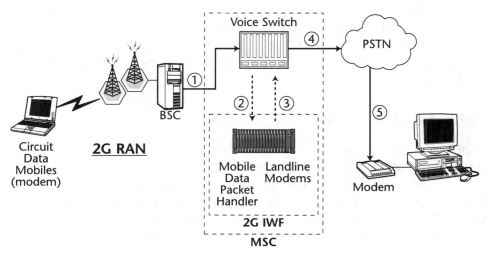

**Figure 4.2** CDMA circuit-switched data architecture.

One of the popular commercial enhancements to cdmaOne data system is Quick Net Connect (QNC).[1] This technology, originally designed by 3COM, Qualcomm, and Unwired Planet (now renamed to OpenWave systems), to allow carriers to provide users with direct access to the Internet and private IP networks, completely bypassing the PSTN and avoiding the need to use modem dial-out procedure. In QNC the IWF routes data from the CDMA wireless network to the Internet or intranet using a PPTP or L2TP (see Chapter 2 for more information on these tunneling techniques). This avoids the use of PSTN network resources, PSTN setup time, and modem training time, resulting in faster and more reliable end-to-end connection.

### GSM and UMTS Circuit-Switched Data

GSM was originally defined to offer digital speech services, since it was designed at the time when the transition to digital speech was recognized as a necessary improvement to the existing 1G systems promising to increase their capacity, voice quality, and confidentiality. GSM deployment was also considered an essential step toward easier international roaming. Although initially the GSM system did not encompass data access, as computer network access became increasingly important to customers as a new service, and to operators as a new source of revenue, new system capabilities were defined. At this time connection-oriented CSD services as well as connectionless data services such as SMS and Unstructured Supplementary Services Data (USSD—which now sometimes is used to interact with Wireless Application Protocol, or WAP, gateways) were defined.

In GSM, the data call is most often initiated by the GSM subscriber, who establishes a dial-up connection by placing a call to a desired data access number provided by an ISP or a private network. Initially the data call originated by the MS is terminated at an IWF, which then converts the incoming data into regular analog modem format and conducts a dial-up procedure to establish a link over the PSTN to the final point-of-call destination, that is, a remote access server (RAS). PPP normally provides the end-to-end link layer service between the MS and the RAS.

The data received by the IWF is converted to the analog frequency-shift keying (FSK) tones characteristic of analog modems. The resulting end-to-end data call can be viewed as two independent calls: the mobile data path and the PSTN path. The mobile data path refers to the connection between the mobile and the IWF. The PSTN path refers to the connection between the IWF and the remote access server at the call destination.

---

[1] Quick Net Connect is only available for the call originating from the mobile device.

An alternative to this architecture is the Direct Internet Access feature, which is similar to CDMA QNC. With this feature the PPP link that originated at the mobile is terminated at the IWF and then IP packets are routed directly to the Internet. On the downside, the mobile user is limited to the Internet access and other data services provided only by the wireless carrier. This drawback, however, can be avoided by relaying the PPP frames received by the IWF over L2TP, thus enabling the wireless carrier to provide access service to remote networks, by using an L2TP VPN-based approach.

GSM circuit-switched users are offered approximately 9.6 Kbps or 14.4 Kbps of throughput. Actual throughput normally depends on the radio channel conditions and the possible compression protocol negotiated at PPP session setup.

With regard to CSD there is no architectural difference between UMTS and GSM, other than the UMTS ability to provide higher data rate CSD services through the use of more efficient radio interface technology. GSM in turn defined High-Speed Circuit-Switched Data (HSCSD),[2] which we discussed in Chapter 3 and whose capability is described in [GSM TS 02.34]. HSCSD allows for the concurrent use of multiple full-rate traffic channels (TCH/F is a 16-Kbps channel) to achieve greater data rates. The HSCSD MS can also request asymmetric bearers.

An MS could request transparent transfer of data over the wireless network circuits. In this case the GSM/UMTS network is simply providing the user with a channel that delivers bits from the MS to the external network, and from the external network to the MS. This kind of service is normally not appropriate for the transfer of TCP/IP-based applications data because of the relatively high bit error rate that would affect communication to the point where TCP/IP throughput would be very low and, at times, close to zero. Instead, this service may be useful for unrestricted delivery of bit streams.

### GSM/UMTS CSD Service Capabilities

The GSM and UMTS *CS domain* offers a set of basic bearer services listed below [GSM TS 02.02]:

**UDI.** This service provides transfer of unrestricted digital information.

**3.1 kHz.** Service used to select a 3.1 kHz audio interworking function at the MSC, this service category is used when interworking with the ISDN or PSTN 3.1-kHz audio service and includes the capability to select a modem at the interworking function. "External to the PLMN" indicates that the 3.1 kHz audio service is only used outside of the

---

[2] Note that HSCSD is also defined in UMTS framework.

PLMN, in the ISDN/PSTN. The connection within the PLMN, user access point to the interworking function, is an unrestricted digital connection.

**PAD.** Through this service, access to a PAD (for X.25-based PSPDN) is provided.

**Packet.** This service provides direct interworking to a packet or ISDN network (normally X.25, X.31, and so on).

**Alternate speech and data.** This service makes possible swapping between voice and data during the duration of the call.

**Speech followed by data.** With this feature, it is possible to start a call in speech mode and then to switch to data, but not the other way around.

Of these, we will be looking only at the UDI and 3.1-kHz services. In most cases, the service is based on the 3.1-kHz bearer, since we are considering PPP-based access for the purposes of advanced data services. UDI and 3.1-kHz service are the most widely used at this time, so we will focus on them.

3.1-kHz and UDI bearer services are very much related. In fact, the *3.1 kHz* term does not apply to the PLMN, rather it represents the PSTN resources used to provide end-to-end connectivity (the spectrum required to transmit a voice conversation) that a modem at an IWF can use to connect to the data network accessed by the PLMN subscribers. Typically UDI and 3.1 kHz use the same physical bearer from mobile to IWF, and different bearers after IWF toward the external (to the PLMN) network. UDI is normally used to support ISDN-type connectivity, while 3.1 kHz is used for analog connectivity to the external networks. Typically, V.110- and V.120-compliant terminal adaptation is used for ISDN connectivity. Using UDI with V.110 or V.120 is also possible to provide direct digital access to data networks, thus achieving faster connection setup.

UDI and 3.1 kHz services are characterized by a very important QoS parameter partially covered earlier in the chapter: the transparent or nontransparent types of the bearer service. As mentioned, a service is transparent if the network behaves simply as a bit pipe, totally unaware of what user protocol is being transmitted. In nontransparent mode the MS and the IWF are the termination point of the Radio Link Protocol. This protocol provides a set of services to the user protocol relay functionality. The user protocol (PPP for common network access services) is being encapsulated in RLP frames. These frames are normally numbered and smaller than the user protocol, so that the retransmission unit size is normally between one and two orders of magnitude smaller than the original user protocol packet or frame size.

This allows retransmission of fewer octets over the air when a part of the user data gets corrupted because of bad radio conditions. The numbered frames can get retransmitted—via ARQ—and also sequence numbers can be used to implement flow control mechanisms, so that when there is excess incoming traffic that the radio interface cannot handle, the RLP sends backpressure signals to the relay functionality (at the terminal side and at the IWF), so that user traffic flow can be paced appropriately. The RLP enhances the link quality from a bit error rate (BER) point of view; that is, it provides some degree of independence of the BER from the radio fading conditions. Though this independence may have some throughput costs, it is normally better than what a transparent service not attempting retransmissions over the radio interface would achieve.

The rest of this chapter focuses on packet-switched data, since we believe it is more likely to be found in future networks and will provide more efficient foundation for delivery of advanced data services such as Mobile VPN.

## CDMA2000 Packet Data

In this section we describe the core packet data architecture associated with the CDMA2000 radio interface. This architecture is described in 3GPP2 recommendations and TIA standards such as [IS835] and [TS115]. It allows CDMA2000 cellular wireless service providers to offer bidirectional packet data services using the Internet Protocol. To provide this functionality, CDMA2000 utilizes two access methods: Simple IP and Mobile IP.

In *Simple IP*, the service provider must assign the user a dynamic IP address. This address stays constant while the user maintains connection with the same IP network within a wireless carrier's domain—that is, until the user does not exit the coverage area of the same Packet Data Serving Node (PDSN). A new IP address must, however, be obtained when the user moves into a geographical area attached to a different IP network—that is, into the coverage area of another PDSN. Simple IP service does not include any tunneling scheme providing mobility on a network layer described in the beginning of this chapter and supports mobility only within certain geographical boundaries.

**NOTE** One of the significant advantages of Simple IP lies in the fact that unlike Mobile IP it does not require special software of any kind to be installed in the mobile station. All the MS needs is the CDMA2000 terminal capabilities and a standard PPP stack similar to that used to establish wireline dial-up session, usually bundled with most modern operating systems such as PocketPC2002 and Windows XP.

The *Mobile IP* access method is mostly based on [RFC2002] now super-seded by [RFC3220], described in Chapter 2. The mobile station is first attached to serving PDSN, supporting FA functionality, and assigned an IP address by its Home Agent (HA). Mobile IP enables a mobile station to maintain its IP address for the duration of a session while moving through CDMA2000 or other systems supporting Mobile IP.

For mobile stations compatible with a TIA/EIA [IS-2000] standard attached to a CDMA2000-1x network, available data rate can vary between the fundamental rate of 9.6 Kbps and any of the following burst rates:[3]

- 19.2  Kbps
- 38.4  Kbps
- 76.8  Kbps
- 153.6 Kbps

These higher-speed bursts are allocated by the infrastructure based on user need (data backlog in either direction), and resource availability (both airlink bandwidth and infrastructure elements). Bursts are typically allocated to a given mobile for a short duration of time of 1 to 2 seconds. The resource and mobile situation is then reevaluated. Burst allocation is performed independently on the forward and reverse channels.

## CDMA2000 Packet Data Architecture

The architecture of CDMA2000 data system is based on the following components (as shown in Figure 4.3):

- A mobile station in a form of a handset, PDA, or PCMCIA card in handheld/portable computer supporting Simple IP or Mobile IP client or both
- CDMA2000-1x Radio Access Network (RAN)
- Packet Control Function (PCF)
- Packet Data Serving Node (PDSN) supporting FA functionality in case of Mobile IP
- Home and foreign AAA servers
- Home Agent (for the Mobile IP access method)

---

[3] There is also a parallel set of burst rates defined in the CDMA2000 radio link standards based on multiples of 14.4 instead of 9.6. However, these rates are not normally implemented because the resulting top rate of 230.4 Kbps does not have good spatial coverage for most real-world RF environments, making this rate-set family less attractive in commercial networks than the 9.6 family.

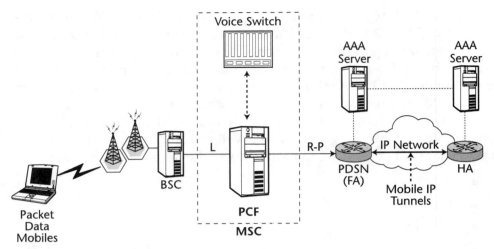

**Figure 4.3**    Example of CDMA2000 packet data architecture.

When the mobile station connects to the CDMA2000 base station, it first establishes a connection to a PDSN. In the case of Mobile IP, the mobile station is then connected to its serving HA by a tunnel between PDSN/FA and the HA established using Mobile IP. The IP address of the mobile station is assigned from the address space of its *Home* network, either statically provisioned or dynamically allocated by the HA at the beginning of the session. On a high-level Mobile IP authentication and authorization is normally performed by both the PDSN and HA by querying the AAA infrastructure (more on this in Chapter 7). In the case of Simple IP, the address must be assigned to mobile station by the PDSN and cannot be statically provisioned in the MS. The authentication for this access method is based only on PDSN.

The connection between the mobile station and its serving PDSN requires a second layer of connectivity to be established for successful IP communication. This connectivity is provided by Point-to-Point Protocol (PPP) as defined by [RFC1661] and supporting IPCP, LCP, PAP, and Challenge Handshake Authentication Protocol (CHAP).[4] PPP is initiated by the mobile station during connection negotiation and is terminated by the PDSN. Between the CDMA2000 radio network and PDSN, PPP traffic is encapsulated into the Radio-Packet (R-P) interface. CDMA2000 protocol stack examples for both the Simple IP and Mobile IP cases are shown in Figure 4.4.

---

[4] Note that CHAP support is not required for Mobile IP and is optional for Simple IP access methods.

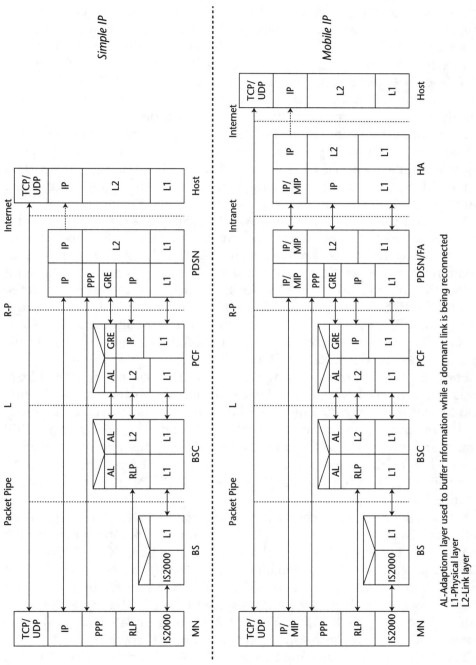

AL-Adaptionn layer used to buffer information while a dormant link is being reconnected
L1-Physical layer
L2-Link layer

**Figure 4.4** Examples of CDMA2000 data service protocol stacks.

The *PCF* shown in this figure is the element of the CDMA2000 Radio Access Network responsible for R-P interface setup and processing. It is often implemented as a component of CDMA2000 MSC. One exception is CDMA2000-1xEV-DO architecture, which does not rely on MSC. The PCF can be implemented there as a part of 1xEV Radio Network Controller (RNC—or BSC depending on the vendor). Stand-alone PCF implementation is also possible. Once link layer connections are established, the PCF simply relays PPP frames between the mobile device and the PDSN. Another important function of PCF is providing micro-mobility support, which is accomplished by allowing the MS to change the PCF while keeping the mobile anchored on the same PDSN and buffering the user data while a dormant radio link is being re-connected. The significance of the latter feature is explained later in the chapter.

The major role of *PDSN* in CDMA2000 architecture is to terminate PPP sessions originated from the mobile station and provide FA functionality, in case Mobile IP service is requested, or to deliver IP packets to the appropriate next hop when Simple IP is used. The PDSN is also charged with authenticating the users and authorizing them for requested services. Finally the PDSN is responsible for establishing, maintaining, and terminating the PPP-based link layer connection to the mobile station. Optionally, PDSN must support secure reverse tunneling to the Home Agent (this option is described in detail in Chapter 7).

For basic Internet service using the Simple IP access method, the PDSN assigns a dynamic IP address to the mobile, terminates the user's PPP link, and forwards packets directly toward the Internet via the default gateway router on the service provider backbone IP network. The normal PPP timers are enforced, and the packets from the mobile may be checked to ensure the mobile is using the source IP address assigned by the PDSN. (Among other filtering rules and policies, the PDSN may implement in Simple IP mode.)

For Mobile IP access methods, the PDSN establishes the Mobile IP protocol connectivity to the mobile station's home network represented by the HA, which is responsible for IP address assignment. The PDSN must support an AAA client functionality to aid in partial authentication of the mobile by local AAA server. Per [IS835], the PDSN is also required to support Van Jacobson TCP/IP header compression and three PPP compression algorithms: Stac LZS [RFC1974], MPPC [RFC2118], and Deflate [RFC2394]—the latter mostly used by Linux- and UNIX-based mobile stations.

The R-P interface connecting PCF and PDSN—also defined by TIA/EIA as A10/A11—is an open interface based on the GRE Tunneling Protocol and is used to connect radio network and PDSN. The R-P interface protocol is actually similar to the Mobile IP where the PCF acts as the FA and the PDSN acts as the HA (the R-P interface uses GRE tunnels for the traffic plane and Mobile IP-like RRQ/RRP messages for signaling). There are a few reasons for the introduction of R-P interface or in other words "splitting" PCF and PDSN functionalities. By supporting the R-P interface, IP-based mobile devices can cross MSC boundaries without impacting the continuity of user sessions. In other words, if the user moves to another MSC coverage area, the user session is not disconnected and the user is not forced to reconnect via the new MSC and obtain a new IP address. This is accomplished by performing *PCF transfers* while keeping mobile devices anchored to the same PDSN. This does, however, require that all serving PCFs have network connections to the same pool of PDSNs (see the case study in Chapter 7). Another purpose of splitting PDSN from PCF is to allow service providers to select PDSNs from third-party vendors, other than those proving the bulk of their infrastructure including MSCs and PCFs. R-P therefore enables wireless carriers to introduce multivendor PDSN solutions into their network. Not surprisingly, the carrier community was the most vocal during the R-P standardization process.

## Mobile Station Perspective

The CDMA2000 mobile station can authenticate with the service provider's HLR for wireless access and authenticate with the PDSN and HA, using the Simple IP or Mobile IP access methods, for data network access. The mobile stations are required to support a standard PPP networking protocol and be capable of supporting CHAP-based authentication during PPP authentication phase for Simple IP service. For Mobile IP service, the mobile device must also support the Mobile IP client as described by the [IS-835]. In this mode, the mobile station communicates with its Home Agent via serving PDSN in the visited network. If the mobile supports one or more of the optional PPP compression algorithm options such as MPPC or Stac LZS, then PPP compression during the connection phase with the PDSN can be negotiated, thus optimizing radio network resources usage and enhancing the user experience via a higher effective data rate.

### *Dormancy*

The mobile device is expected to support airlink "dormancy" (as defined by TIA [IS-707A1]), which allows either the mobile or the MSC to time out the active airlink connection after a period of inactivity and to release the air interface and serving base station resources. If either the mobile station or the associated PCF have packets to send while dormant, the connection is reactivated and the transmission continues. Dormant mobile stations are defined as stations that do not have an active link layer connection to the serving PCF. All mobile stations—active and dormant—registered using Mobile IP access method have an entry in the PDSN visitor list and a binding with the corresponding HA.[5]

The PDSN serving the users on the foreign network serves as the default router for all registered mobile users, active and dormant, and maintains host routes to them. For Mobile IP mode the PDSN/FA keeps track of the time remaining of the *registration lifetime* for each mobile station in its routing tables and the MS is responsible for renewing its lifetime with the HA. If the mobile does not re-register before the expiry of the registration lifetime, the PDSN will close the link with the PCF for this mobile and terminate the mobile's session (and the HA will do likewise if the mobile has not re-registered via some other PDSN). Once the mobile station's registration lifetime has expired, the PDSN/FA will stop routing packets to it. To receive and send packets, dormant stations must therefore transition to the active state. Given that any registered mobile stations at any moment can be in active or dormant sub-states, the PDSN generally does not require an indication of the state of PPP links to mobile stations except for the current dormancy timer value for that particular link. Traffic may arrive on the dormant link at any time, forcing the associated mobile station to transition to active state. For active, traffic-carrying PPP links, the PDSN terminates the PPP session with the mobile station and relays the encapsulated IP traffic to the mobile from the HA or from the mobile to the HA via reverse tunneling. A separate tunnel exists for each unique HA for all registered users.

### Mobile Station Types

There are two basic types of mobile station configurations—relay model and network model. In *relay model* mobile stations, the CDMA2000 Mobile Terminal is connected to another portable data terminal device such as a laptop, handheld computing device, or some other embedded data terminal. The relay model phone does not terminate any of the protocol layers except for the CDMA2000 physical layer (radio interface) and RLP layers. The attached data terminal device must terminate all other higher-layer protocols (PPP, IP, TCP/UDP, etc.).

*Network model* mobile stations, in addition to the radio interface, terminate all necessary protocols and do not require any additional data terminal devices. The mobile phone itself provides all user input and display

---

[5] Note that dormancy only affects the radio link level connection, while all dormant mobiles retain a PPP link with the PDSN which is not aware of the active/dormant state of the MN.

capabilities—as well as a user applications—to make use of the packet data network. Examples of this kind of phone include the "smart phone" or "micro-browser" phone. These devices normally include some embedded Web browsing or information service application, as well as a display screen for viewing the information retrieved from the Internet server. Such kind of terminals may also offer the ability to connect a laptop to a data network via a PPP connection terminated at the terminal itself. In this configuration, the phone can support embedded applications such as a micro-browser and also allow general-purpose use by the external data endpoint.

## CDMA2000 Mobility Levels

CDMA2000 packet data architecture defines as many as three levels of mobility for the mobile station, as depicted in Figure 4.5. One level is represented at the physical layer by BTS-to-BTS soft or semisoft handoff, while the mobile station is anchored at the same PCF. This is accomplished by the CDMA2000 radio access and is invisible to both PCF and PDSN.

The second mobility level is represented by the R-P interface on the link layer, which allows for a transparent handoff from PCF to PCF while keeping the session at the same PDSN. In this case, two options described previously come into play: dormant and active. In active state, when the user crosses PCF boundary, the handoff is transparent for the mobile station. The MS participates in a semisoft handoff to the new BSC (or MSC, depending on the vendor), while the link layer data session remains anchored to the original PCF for the duration of the call and the mobile is in the active state. In other words, when the mobile station is in the active state, change of serving PCF will not occur.

When a mobile crosses a PCF coverage boundary while dormant, the mobile will trigger reactivation at a new BSC (MSC) to establish a new PCF connection. That results in a PCF but not necessarily a PDSN change if both the current and previous PCFs were attached to the same PDSN. The new PCF attempts to assign the mobile to its current serving PDSN. If the new PCF has connectivity to that PDSN, the PPP session previously established between the mobile station and the PDSN will be totally unaffected.

The third level of mobility, the network layer, is the inter-PDSN handoff, based on the use of Mobile IP protocol. Let's assume that the mobile station has registered with the HA and PDSN (the MS has been authenticated by each of them) to establish the Mobile IP tunnel over which traffic is delivered. Whenever the mobile roams to a location that is served by a PCF connected to a different PDSN, the mobile receives an indication that it must reregister with this new PDSN. This reregistration updates the mobility binding tables at the HA, so that all subsequent traffic is routed to the new

PDSN for this mobile. In this case the mobile's PPP link is impacted by this change while the IP layer stays intact, and the mobility remains invisible to the mobile station's correspondents.

Note that the last type of handoff is not available in Simple IP mode; Simple IP provides only partial mobility, via the other two levels, to the mobile station. One of the functions of the R-P interface is to bring Simple IP service closer in functionality to Mobile IP service, along with addressing other problems. For example, it addresses the situations where the mobile station changes its point of attachment to the network so frequently that basic Mobile IP tunnel establishment introduces significant network overhead in terms of the increased signaling messages. Another often-cited problem is the latency of establishing each new tunnel, which introduces delays or gaps during which user data is unavailable. This delay is inherent in the round-trip incurred by Mobile IP as the registration request is sent to the HA and the response is sent back to the PDSN.

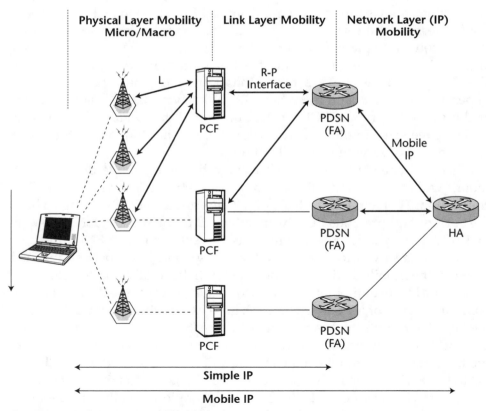

**Figure 4.5**  CDMA2000 mobility hierarchy.

## CDMA2000 Mobile AAA

CDMA2000, just like the majority of other cellular systems, supports the concept of home and visited networks. A CDMA2000 subscriber has an account established with one wireless carrier, which provides the user with wireless voice and data services. This same wireless carrier may provide a *home network* for the mobile subscriber. The home network holds user profile and authentication information. When the user roams into the territory of a different wireless carrier—that is, a visited network—that carrier must obtain the authentication information and the *service profile* for this particular user from its home network. The service profile indicates what radio resources the user is authorized to use, such as a maximum bandwidth or access priority. In CDMA2000 the user profiles are stored in a Home Location Register (HLR) located in the home network and are temporarily retrieved into a Visitor Location Register (VLR) located in the serving network. The HLR and VLR are databases housed on fault-tolerant computing platforms. Similar procedures take place to authenticate the user access to data networks. CDMA2000 packet data architecture, depicted in Figure 4.6, is based on the concept of home and foreign data networks represented by HA and PDSN and home and foreign AAA servers—for example, RADIUS or DIAMETER—as defined by [RFC3141].

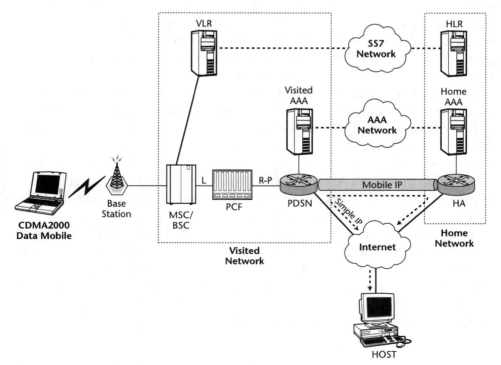

**Figure 4.6**   Typical CDMA2000 core network with AAA subsystems.

Mobile stations requesting data service in CDMA2000 systems will have to be authenticated twice: on the physical layer and on the link layer (or, using other terminology frequently used in the industry, both wireless access and network access authentication will take place). Physical layer (or wireless access and user terminal equipment) authentication is performed by the cellular wireless system's HLR and VLR infrastructure. It is based on an International Mobile Station (IMSI) and is defined in [IS-2000] (the details of this authentication method are outside the scope of this book). CDMA2000 link layer, or packet data network access, authentication of the mobile station is conducted by the infrastructure of AAA servers and clients, the latter being hosted by PDSNs and HAs. It is based on a Network Access Identifier[6] (NAI, defined by IETF [RFC2486])—that is, a user identifier of the form user@homedomain, which allows the visited network to identify the home network AAA server by mapping the "homedomain" label to the home AAA IP address. A challenge from the PDSN also allows for protection from replay-based attacks.

Among other services, NAI allows for distribution of specific Mobile IP security association information to support PDSN/HA authentication during mobile registration, HA assignment, and inter-PDSN handoff. Note that data network AAA authenticates the user, as opposed to physical layer authentication, which only authenticates the mobile. Therefore, users wishing to gain access to public or private data networks are presented with a login and password sequence, familiar to wireline remote data access users, in addition to mobile device authentication taking place at registration stage, which results in the momentary hesitation at phone startup familiar to most mobile phone users.

The CDMA2000 data subsystem provides two user authentication mechanisms when simple IP or Mobile IP access methods are requested, as defined in [IS835] and [RFC3141]. As mentioned, for the Simple IP access mode, authentication is based on CHAP, which is a part of PPP negotiation. In CHAP, the PDSN challenges the mobile station with a random value to which it must respond with a signature based on MD-5 digest of the challenge, a username, and a password. The PDSN passes the challenge/response pair to the home AAA server for user authentication.

For Mobile IP, the PDSN sends a similar challenge within the agent advertisement message to the mobile station. Again, the MS must respond to the challenge with a signature and NAI that is verified by the home network, but this time the response is sent along with the Mobile IP

---

[6] More details on NAI and other Mobile IP extensions are provided in Appendix A.

registration request rather than during PPP session establishment. Both of these mechanisms rely on shared secrets associated with the NAI, which are stored in the home network, and both will be supported by the same AAA infrastructure. In both cases the accounting data is collected in the PDSN and transferred to the AAA server. The PDSN collects data usage statistics for each user, combines these with the radio access accounting records sent by the PCF, and forwards them to the local AAA server. Note that accounting information is collected by both the PCF and the PDSN. For roaming users, the AAA server may be configured to forward a copy of all RADIUS accounting records to the home AAA server in addition to keeping a copy at the visited AAA server.

When a handoff between two PDSNs occurs, an *Accounting Stop* message is sent to the AAA server from the releasing PDSN, and an *Accounting Start* is sent to the AAA server from the connecting PDSN (more details on this and other accounting mechanisms is provided in Appendix B). The Accounting Stop from the releasing PDSN may arrive some time after the Accounting Start from the new PDSN (the releasing PDSN may not be aware that mobile has moved, but it must wait for a Registration Lifetime or PPP Inactivity timeout to end the session). This means the billing server must accept multiple stop/start sequences from different PDSNs that contain overlap and treat these as a single session, per [IS 835]. When a PPP inactivity timer or an MIP lifetime expires, or the mobile terminates the session, the R-P link is released and an Accounting Stop is sent to the AAA server.

Having so far discussed the details of packet data services in CDMA2000, in the next section, we give the same level of exposure to the UMTS and GPRS systems.

## GSM and UMTS Packet Data: General Packet Radio Service and UMTS PS Domain

In this section we describe the GPRS and UMTS PS domain systems and document the services offered to the MS. We look at the possible ways an MS can access packet data networks, such as which protocols can be used and in which way user authentication can take place. Also, we outline what the current status of the market is, relative to some capabilities we believe will be adding value to service providers but perhaps are not yet widely offered by most network equipment manufacturers. Finally, we conclude the section with a discussion of services capabilities of the two systems.

## GPRS Elements

The GPRS system is a *packet* data extension of the GSM system, which was originally designed for *circuit* services. GPRS enables the support of *packet-*based data transmission over the radio interface and packet data mobility within the core network. Deploying GPRS entails a Base Station Subsystems (BSS) software-only upgrade, which allows multiplexing of data services over the slots not occupied by speech services, flow control, and retransmission mechanisms necessary to deliver data services over the (GSM) radio transmission technology. It also requires updating the HLR software and installing new core network nodes: the Serving GPRS Support Node and Gateway GPRS Support Node (SGSN and GGSN). DNS, address management system, AAA, billing, and intelligent network are additional components that are parts of advanced GPRS services. The GPRS architecture and specifications were defined by ETSI and now are maintained by 3GPP (see Chapter 1). The GPRS architecture is depicted in Figure 4.7.

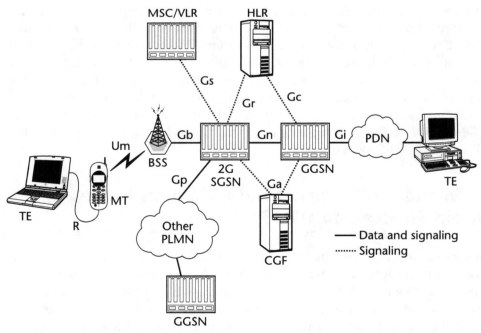

**Figure 4.7** GPRS architecture.

The SGSN in GPRS, also referred to as 2G SGSN, offers network layer compression services, segmentation and reassembly functionality, logical link layer framing and multiplexing, ciphering, as well as handling MS signaling and mobility management (within the BSS, between SGSNs), and managing GTP tunnels established toward GGSNs. The SGSN also interacts with the HLR and the intelligent network, the MSC, and the SMS Service Center (SMS-SC).

The GGSN "anchors" the data communications session and provides access to packet data networks by supporting the termination of GTP tunnels from the SGSN to which the MS is currently attached. GGSNs are also used in IP networks to provide a foundation and a gateway to advanced packet data services such as Web browsing, WAP; remote private networks, including networks used to provide mobility support for nonroaming users (intra-PLMN network), roaming (GPRS Roaming Exchange or GRX network), and GPRS network element functions.

## UMTS Elements

The UMTS offers packet data services over its *PS domain* (see the next section). Its architecture, shown in Figure 4.8, is similar to the GPRS architecture. The same network nodes are involved in the core network, although the GTP protocol used in UMTS (GTPv1) is not backward-compatible with the GTP protocol versions used for GPRS (GTPv0). Other differences are in the service capabilities—for instance, UMTS supports its own QoS framework along with more multimedia capabilities. Also, the UMTS SGSN, also referred to as 3G SGSN, does not terminate any link layer protocols nor provide network layer compression or ciphering. Instead, it simply relays packets between GGSNs and Radio Network Controllers over GTP tunnels. (As mentioned in Chapter 3, RNCs are somewhat similar in functionality to BSCs in GSM.) In UMTS, the RNC function is managing mobility of MSs among the Node Bs (UMTS base stations), which it controls transparently to the UMTS PS domain. According to one of the UMTS fundamental principles, the RAN details must be hidden from the core. This is one of the reasons why all the UMTS link layer functionality has been moved from SGSN to the RNC.

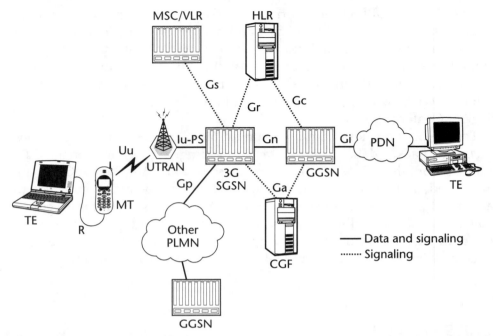

**Figure 4.8**    3GPP UMTS architecture.

The system specification for both the GPRS and UMTS PS domain for Release 99, and later releases, is contained in one single document—[3GPP TS 23.060]. It defines all the system-level aspects from 3GPP R99 onward, and it points out the differences between the two systems. Such differences are normally only minor or are related to the handling of terminal mobility management and to interfacing the RAN. In releases before R99, GPRS architecture was specified by [GSM TS 03.60].

## GPRS and UMTS PS Domain System Architecture

The GPRS system defines mainly two components: the BSS and the PLMN backbone. The GPRS BSS is the GSM BSS augmented with a Packet Control Unit (PCU) that is used to upgrade the GSM BSS to support packet services. The PLMN backbone includes two new nodes introduced previously: the SGSN and the GGSN. The GGSN and the SGSN are connected via an IP network and interact via the Gn interface, based on the GTP protocol.

When a user roams, that user attaches to a SGSN in the visited network and a GGSN in either the home network or the visited network. If the GGSN is in the home network, the IP network used to connect the visited SGSN to the home GGSN is named the *Inter-PLMN Backbone Network*. The Inter-PLMN Backbone Network is usually offered by network service providers under the name GPRS Roaming Exchange (GRX), which was defined by the GSM association. Candidates to operate the GRX network have to comply with requirements set forth by the GSM Association. A curious aspect of GPRS as it relates to GRX is that the SGSN in the visited network and the GGSN in the home network interact over GRX network via an interface, called Gp, which is entirely identical from a protocol perspective to the Gn interface. However, since wireless network providers may decide to provide additional security mechanisms to protect inter-PLMN traffic—and will have to do so starting from 3GPP Release 5—it was determined that a new name for the interface was in order, even though such security mechanisms were not standardized.

The UMTS system takes into account from the outset the need of supporting both a packet-switched (PS) core network and a circuit-switched (CS) core network. One of the principles for general 3G access network specifications has been independence of the core network from the radio interface and support of multiple cores. 3GPP has defined a *CS domain* for circuit services and a PS domain for packet-based services. Since mobility within the UMTS Terrestrial Radio Access Network (UTRAN) must be transparent to the UMTS core, the later should be unaware that the MS would be camped on a particular base station. To stress this need, 3GPP decided that the RAN would be specified as a network made of RNCs and Node Bs, moving away from GSM terminology. Also, early models and research projects introduced the concept of Access Network Operators (ANOs) and of Core Network Operators (CNOs). These concepts did not make it into the official terminology, and standardization efforts did not address their support as officially recognized entities.

The UMTS PS domain core is identical to the GPRS core. In fact, starting from R99, both systems' specifications have no technical differences as far as the core network is concerned. Notably, however, there are some significant differences between the available services in GPRS R98 and GPRS/UMTS R99 that will be highlighted in the remainder of this chapter. We think it is useful remarking the differences between the two systems, because GPRS will be normally identified with R98, since the vast majority of networks will be supporting UMTS only from R99 onward (with the exception of some North American operators who might elect to support EDGE access network, for reasons discussed in Chapter 3).

GPRS specifications define new protocol layers in the BSS—Radio Link Control (RLC) and Medium Access Control (MAC)—that enable the usage of the existing GSM framing structure for GPRS. The GSM system specification allows for the usage of one to eight time slots in both uplink and downlink. The standard specs define the way to adapt different network protocols to the logical link service offered by this system—Sub-Network Dependent Convergence Protocol (SNDCP)—by relaying Logical Link Control (LLC) frames between the MS and the SGSN. This relay functionality is provided by the PCU, which enables enhancements of the GSM BSS to become GPRS services-capable.

The GSM RAN, augmented with the PCU functionality, is connected to the GPRS core via the Gb interface, which defines a Frame Relay-based network service on top of which the base station subsystem GPRS protocol (BSSGP) is carried (Figure 4.9). BSSGP is used to support logical channels on top of the Frame Relay network service. These logical channels are used to implement an LLC protocol frame's routing between the BSS and the core, so that the BSC+PCU can route LLC frames to the correct cell. In the uplink, the BSSGP carries the cell ID information, and any LLC frame transported on the BSSGP protocol can offer location information. Typically, an MS updates the location (cell update) when it is with an active session (sending and receiving data) by sending an LLC frame.

SNDCP, LLC, and BSSGP—as well as the Frame Relay-based network service—are terminated at the 2G SGSN. Since both user traffic and control information are transported over these protocols, and the link layer services to the mobile are terminated at the 2G SGSN, a 2G SGSN is particularly complex. This complexity potentially brings severe limitations on its scalability. In part to address this situation, during the UMTS standards development, the decision was taken not to terminate link layers at the SGSN, thus reducing the complexity of this element and enabling it to scale to support many more subscribers and a wider coverage area. Scaling an SGSN to cover wider area is important to minimize mobility management signaling associated to handoff and to allow the handoffs to be, on the average, smoother. In UMTS, 3GPP defined a cleaner interface between the RAN and the PS Domain Core network, named the *Iu-PS* interface (Iu interface in its general form was first introduced in Chapter 3). This interface is based on IP over ATM transport for the user plane (the link layer can be based on any technology starting from releases later than R99, such as R4 and R5), and on the Radio Access Network Application Part (RANAP) protocol for the control plane. RANAP is an SCCP user application. SCCP is carried over a broadband SS7 (MTP3-B, described in [Q.2210] using the service of SAAL-NNI (Signaling ATM Adaptation Layer Network-to-Network Interface) [Q.2100] or an IP transport based on SIGTRAN protocols—namely, M3UA and SCCP.

User packets are transferred over the Iu-PS using GTP/UDP/IP transport, and then relayed by the RNC toward the MS using radio link protocols (PDCP/RLC/MAC). Packed Data Convergence Protocol (PDCP) functionality in UMTS is similar to SNDCP in GPRS. It also supports header compression protocols that are particularly useful for reduction of the overhead for real-time applications—in particular, for future Voice over IP (VoIP) applications that will be characterizing the evolution of the UMTS system as we move to UMTS Release 5. (See Chapter 9 on the system evolution for more details.) The RLC and MAC layers are used to implement the radio link layer. They implement the logical link layer channels over the W-CDMA radio interface physical layer. Figure 4.9 summarizes the discussion on the protocol stacks we have just completed by giving a synopsis of the *user plane* of both the GPRS and UMTS systems.

The Mobile Terminal communicates control information to the GPRS or UMTS core network using a Radio Interface Layer 3 protocol (RIL3) specified in [3GPP TS 24.008]. This includes GPRS (or packet, in UMTS) mobility management and session management. In GPRS, RIL3 is carried over the LLC NULL channel and is handled by the SGSN on the core network side. In UMTS, this protocol is transported using the direct transfer procedure over the RANAP protocol. The SGSN elaborates control information from the terminal, and as a consequence, it can interact with other entities in the network. For example, it can open MAP dialogues with the HLR for security procedures, subscriber information retrieval, or location update purposes. It also can interact with the intelligent network via the CAMEL [3GPP TS 23.078] interface, for instance, to offer prepaid data services to GPRS users.

Packed data sessions in GPRS and UMTS are established by setting up and maintaining GTP tunnels toward GGSNs. A GTP tunnel is an encapsulation of the user packets between the GGSN and the SGSN in GTP/UDP/IP. Actually, the encapsulation was also defined to be GTP/TCP/IP, but from R99 onward it was the common consensus that this was not needed, since its only purpose was transferring X.25 in a reliable manner over the PLMN backbone. As is clear now, X.25 is by no means the future of data networking; the wireless carriers aren't going to offer X.25 service, and the networking equipment vendors won't support it.

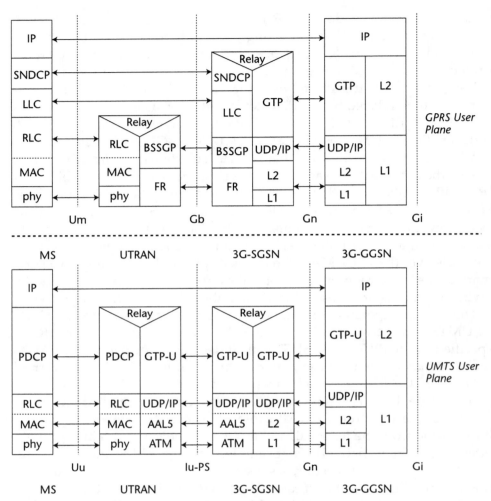

**Figure 4.9** The GPRS user plane and control plane.

Another function of the SGSN is the collection of charging data and communication to a Charging Gateway Function (CGF) over the Ga interface (see Figure 4.7). The Ga interface is based on the GTP protocol [3GPP TS 32.015]. The SGSN can also support SMS services via the Gd interface to SMS-GMSC and interact with the MSC/VLR via the Gs interface for paging coordination and combined location update. Because of the limited scope of this book we will not be addressing these aspects of SGSN in more detail. Please refer to the Bibliography if you wish to find articles and Web sites that contain a more in-depth discussion of this subject.

## GPRS and UMTS PS Domain Service Capabilities

The GPRS and UMTS PS domain systems are in principle multi-protocol and neutral to the nature of the network layer or link layers of the user traffic. The user protocols are also called Packet Data Protocols (PDPs). The type of protocol is identified with the term "PDP type." Initially, the GPRS system supported many more PDP types than were actually implemented and deployed. For example, 3GPP was recommending PDP type supporting X.25 based on TCP/IP-reliable delivery of GTP-encapsulated X.25 packets over the GPRS backbone. This PDP type was later removed from the specs from R99 onward, and it is unlikely that it will be present in any real deployment of earlier releases. Another example is Internet Host Octet Stream Protocol (IHOSS), which was supporting a transparent stream of octets between GGSN and MS. This PDP type was also removed from the R99 specifications.

GPRS provides support for both IPv4 and IPv6, but IPv6 cannot be realistically expected to see deployment until sometime in 2004. It is supporting PPP type PDP starting from R98, but unfortunately, terminal vendors did not rush to implement the support of this PDP type, and we haven't seen any real support of this PDP type as yet. In fact, we have witnessed an unexpected amount of controversy surrounding this PDP type. Some vendors have tried to remove it from the standard, despite the fact that in principle this PDP type would prove to be beneficial in offering advanced data services such as access to private networks (as described in Chapter 6).

The GPRS and UMTS PS domain systems offer an *unreliable* transport channel from the GGSN to the MS. This channel can be characterized by a number of QoS parameters, which are different for releases prior to R99 and releases from R99 onward. In R99 and later releases, it is possible to differentiate the handling of packets belonging to the same user session. This is accomplished by establishing multiple bearers (PDP contexts) of different traffic class and QoS profile, associated with the same session, and by forwarding packets on to the appropriate bearers based on some classification rules established at the GGSN and at the MS. This capability has been introduced because of the requirement of offering multimedia capabilities via the UMTS system, at the basis of the service aspects expected from 3G cellular systems. In R98, instead, only one level of QoS (and only one PDP context) can be associated to a session.

## GPRS and UMTS PS Domain Terminal

There are three different classes of GPRS MS:

**Class A.** This class allows for concurrent support of GSM and GPRS services.

**Class B.** In this class, the MS monitors both GSM and GPRS paging channels, but only one service can be supported at a time.

**Class C.** In this class, the MS only supports GPRS service (as a data-only device).

Class A terminals require two separate GSM transceivers operating on different frequencies, with independent packet and circuit services. Because of the complexity of such terminals, in R99, operators and manufacturers targeted for the definition of terminals supporting Dual Transfer Mode (DTM). From a service perspective, this approach still offers the simultaneous coexistence of GPRS and GSM voice services like in a Class A terminal, but from a paging perspective they are more similar to a Class B device and require a single transceiver. This approach is made possible via an upgrade of the BSS to route paging over the same channel as used for GPRS, when the MS is GPRS-attached. Note that this enables the paging coordination to happen in the BSS, much like in UMTS systems.

A Mobile Terminal capable of UMTS PS or GPRS access can either be:

- An integrated device offering computing and wireless data access in one physical unit
- Composed of two components: one devoted to wireless access and the other a data applications capable device

The latter configuration is similar to today's laptops, which are equipped with PCMCIA modem cards or a serial connection to a modem. In fact, the 3GPP standards [3GPP TS 29.061], [3GPP TS 27.060] define the existence of two logical components of the MS: the Terminal Equipment (TE) and the Mobile Termination (MT). The TE is the computing-capable part of the MS. The MT is the part devoted to the support of the wireless data access capabilities. When the TE and MT are implemented as separate physical entities, they can be connected by multiple technologies (serial, infrared, Bluetooth, etc.) with the link layer based on PPP, or another proprietary interface. Figures 4.7 and 4.8 show the two MS components as separated by the R interface. Indeed, this interface may be internal between components of a single physical package, rather than between physically distinct entities.

When the PPP interface is used between the TE and MT, the IP PDP type can still be used, but the authentication and configuration material is not transported between the GGSN and the MT using PPP. Rather, it is encapsulated in a Protocol Configuration Options Information Element (PCO IE), transparently relayed between RIL3 and GTP by the SGSN. This way, PPP exists between the TE and MT, but not between the MT and the GGSN. This access mode is called *nontransparent IP access* and is discussed in more depth in Chapter 6.

Mobile Terminals also often come with dual-mode GPRS/GSM and UMTS capability, since many operators will roll out UMTS starting from densely populated and capacity-strained areas, and rely on GPRS for the handling of users in areas where only 2G coverage exists. In fact, this will be one of the most commonly requested terminal features in the early days of UMTS.

## Summary

In this chapter we discussed the support of circuit-switched data and packet-switched data in cellular wireless systems. We addressed both terminal and core network aspects. This, together with the cellular and wireless LAN radio access background provided in Chapter 3, should provide the knowledge necessary to proceed to the discussion of the support of advanced wireless data services in the mobile environment. The next three chapters will be devoted to this topic.

PART

# Two

# MVPN and Advanced Wireless Data Services

# Mobile VPN Fundamentals

VPNs have been widely deployed by the networking industry in various shapes and forms for many years. Lately, VPN has been introduced to IP-based data communications as well. Industry analysts forecast significant growth for the IP VPN industry; for instance, Infonetics Research believes worldwide end-user VPN expenditures are set to grow 275 percent, from $12.8 billion to $46 billion, between 2001 and 2006. It is also expected that VPN will have a significant impact on wireless communications. The latest application of VPN to mobile communications, MVPN—still in its infancy—has many unresolved issues, both from a technical and business perspective. However, its technical framework is already largely defined, and early deployments of its various forms have been undertaken.

In this chapter we introduce MVPN concepts and analyze its technology by first addressing architectures and taxonomy of traditional data VPNs and then moving on to add mobility to the picture. The discussion begins with a VPN definition and an analysis of private networking. We continue this discussion with sections on VPN-enabling technologies and on VPN taxonomy. The chapter concludes with an analysis of wireless versus mobile terminology and the introduction of VPN in mobile environments.

## Defining VPN

Let's expand on the VPN definition we provided back in Chapter 1. VPN combines two concepts: virtual networking and private networking. In a *virtual* network, geographically distributed and remote nodes can interact with each other the way they do in a network where the nodes are collocated. The topology of the virtual network is independent of the physical topology of the facilities used to support it. A casual user of the virtual network, not aware of the physical network setup, would only be able to detect the topology of the virtual network. A virtual network is also managed as a single administrative entity.

*Private* networks are usually defined as nonshared networking facilities combining hosts and clients that belong to the same administrative entity. A good example of a private network is a corporate intranet, which can only be used by a certain number of authorized individuals belonging to that particular corporation. Virtual private networking, thus, is the emulation of private secure data networks over public shared insecure telecommunications facilities (recall the MVPN definition in Chapter 1).

VPN properties include mechanisms for data protection and establishing trust among hosts in virtual networks and incorporation of various methods to enforce and maintain service level agreements (SLAs) and quality of service (QoS) for all entities that make up a Virtual Private Network. VPN can be defined from many perspectives. The preceding definition looks at VPN from a networking standpoint, which should serve us well for the purposes of this book.

In the past few years there have been various attempts to come up with a broader VPN definition that would include applications-level technologies such as TLS (see Chapter 2 for analysis of TLS versus VPN). Other sources deservedly argue that vendors and information technology communities took an overly simplified approach to the matter and turned the whole VPN concept into a generic networking term that is now widely exploited for marketing purposes. While acknowledging the validity of both points of view, especially in light of relative VPN newness, we still see a need for a common working definition for data VPN to transition to practical aspects of its implementation.

## VPN Building Blocks

In this section we provide an overview of the technologies upon which a VPN is built. They include:

- Access control

## COMMON MISCONCEPTIONS ABOUT PRIVATE NETWORKS

Private data networks are still perceived as premium transport technology in comparison to IP VPN, or even ATM and Frame Relay networks. Such a perception is rather deeply rooted (perhaps due in no small part to concerted, almost century-old marketing efforts of telecommunications service providers, insisting on rock-solid security and robustness of their copper and fiber) but often quite unfounded. The most popular examples of such private networks are PSTN, private data leased lines, and dial-up networks in both wireline and wireless environments. In the following paragraph we demonstrate that networks based on dedicated private facilities suffer from the same operational problems as Virtual Private Networks provisioned over shared networking infrastructures.

### PRIVATE NETWORKS ARE SECURE

It is often said that a system is only as secure as its weakest link. Under this credo, a service provider's private networking facilities are only as secure as the individual elements of their networks, such as copper of fiber cables, circuits established over the air link (in the case of wireless communications), and switching and routing equipment on the data path. The weakest link of such a network is usually the physical environment, which cannot altogether be privatized or secured.

For instance, segments of private networks installed in rural and urban underground conduits (conveniently marked by various signs designed to help locate them) or aerial runs (cables on the telephone poles, reachable by almost anyone with proper equipment and protected only by an extra layer of plastic sheathing) are often unguarded and easily accessible not only by maintenance crews but also by unauthorized individuals with possible criminal intent. As a result, the data transmitted between two corporate divisions can be extracted from an underground fiber cable by simply stripping the sheathing and washing off the protecting gel, bending it, and capturing the escaping light with an inexpensive device used by phone company field technicians. The data transmitted over copper loops in metro areas can be accessed even easier by installing bridged taps in unprotected underground conduits or tapping directly into connecting terminals for the facilities routed through apartment building basements in metropolitan areas. Networks based on link layer technologies—such as ATM and Frame Relay—in addition to physical layer vulnerability, are often susceptible to network-based attacks such as malicious intrusion or eavesdropping.

In addition to transport facilities vulnerability, the data transported over private networks is usually not encrypted by either service providers or enterprises. Service providers, for some reason convinced of the impenetrable nature of their cables and underground conduits, do not bother to implement any ciphering schemes or other encryption or authentication measures. Enterprise IT departments, convinced by service providers that the traffic will be carried only over their private (i.e., secure) facilities, in turn see no need to implement complex end-to-end security precautions.

*(continues)*

**COMMON MISCONCEPTIONS ABOUT PRIVATE NETWORKS** *(Continued)*

The situation in cellular or wireless LAN communications is not much better. Access to the information transmitted over the air is even easier than to that transmitted over cables. Thus, early analog systems initially did not provide any data protection or even proper user equipment authentication and usually could be broken into by using inexpensive scanning and decoding equipment. This resulted in massive early analog mobile phone cloning fraud in the 1980s and 1990s.

The wireless industry responded with new digital systems that incorporated air interface ciphering schemes and mobile phone Subscriber Identity Modules—such as SIMs and USIMs—used in GSM and UMTS phones, respectively, to provide user authentication and ciphering. While these measures successfully addressed the mobile devices fraud problems, the issue of data security had not been dealt with equally effectively. The latest 2G and 3G digital cellular systems air interface ciphering schemes protecting the user data are often weak and expensive to implement and often do not have the desired effect on the public's purchasing decisions. Consequently, these systems are made optional in many commercial implementations. Examples include the GSM cipher, which has been broken and consequently updated, and CDMA2000 air interface security considered optional by the TIA standards and often not supported by equipment manufacturers.

PRIVATE NETWORKS ARE ALWAYS RELIABLE

Contrary to popular belief, sometimes leasing a dedicated facility may expose a company to more serious service disruptions and slower recovery than VPN-technology-based networks. Private networks based on dedicated circuits are much less reliable than technologies based on the use of virtual links (such as ATM, Frame Relay, or MPLS networks) or datagram-based solutions (such as IP VPNs), which allow for automatically rerouting the data flow over different virtual paths in case of intermediate link or node failures.

Often, private networks span significant distances over multiple interconnected facilities and traverse multiple types of data equipment, such as switches, routers, cross-connects, and SONET Add/Drop Multiplexers (ADMs). (SONET stands for Synchronous Optical Network.) In the case of a circuit or preprovisioned private line failure, data cannot be automatically rerouted through other paths, and communication is not possible until the service is restored. Incidentally, SONET rings were designed to address this problem, but only at the physical layer (the strongest link of the system) and usually not in end-to-end fashion. It is also rarely the only physical technology used in a private network, exposing an *end-to-end path* to survivability problems.

- Authentication
- Security
- Tunneling
- Service level agreements

These building blocks are common to most types of data VPNs including MVPN. For this reason, their description will help prepare the reader for the detailed discussion of MVPN in Chapters 6 and 7.

## Access Control

Access control in data networking is defined as a set of policies and techniques governing access to private networking resources for authorized parties. Access control mechanisms operate independently of authentication and security and basically define what resources are available for a particular user after the user has been authenticated. In the VPN world the physical entities, such as remote hosts, firewalls, and VPN gateways in the corporate networks involved in communication sessions, are usually responsible for or at least participating in enforcing VPN connection status.
    Examples of decisions include:

- Initiate
- Allow
- Continue
- Reject
- Terminate

The main purpose of any VPN is to allow selective, secure access to remote networking resources. With security and authentication but without access control, the VPN is only protecting transmitted traffic integrity, confidentiality, and preventing unknown users from using the network, but not networking resources. Access control usually depends on information about the entity requesting connection such as identity or credentials—as well as the rules defining access control decision. For example, some VPNs can be governed by a centralized server or other VPN control device located in the service provider's data center, or it can be administered locally by VPN gateway in private networks involved in VPN communication.

The set of rules and actions that defines the access rights to network resources is called the *access control policy*. The access control policy allows for the enforcement of a business goal. For instance, the policy "Allow access to remote access subscribers who have not surpassed 60 hours of usage" can be implemented by using RADIUS-based authentication of the user and incrementing a time counter every time a user gets access. In theory, a RADIUS DISCONNECT message (see [RFC2882]) may be used to interrupt the user session when the 60 hours are surpassed, but sometimes the policy may be enforced only at logon time, trusting users not to permanently stay logged on, or perhaps by putting a session duration limit that would put an upper bound on the usage exceeding the maximum allowed time. Similar policies may be implemented by replacing the time limit with a credit limit that may be associated to a prepaid account.

## Policy Provisioning and Enforcement

The examples of standard-based policy provisioning and enforcement mechanisms include Lightweight Directory Access Protocol (LDAP), a standard for querying a directory, defined by [RFC2820], [RFC2829], and RADIUS. The service definitions and specific access instructions for both mechanisms are stored in centralized databases, accessible by the equipment responsible for access control decisions, such as IP routers, VPN gateways, or intelligent network appliances. The main advantages of these methods are simplicity and ease of management and provisioning. With LDAP, for example, changes in policies can be made by both the service provider and corporate IT administrator by accessing only one server housing a particular database instead of reprovisioning the information stored locally on many network elements.

The use of RADIUS binds the user authentication process with access control and service policy selection. Once the access policy is selected, its enforcement may take place at the device by querying a policy repository using LDAP. In recent years, another protocol named Common Open Policy Service (COPS—[RFC2748]) has been proposed to implement policy decision outsourcing and provisioning for very simple devices. Its use is so far very limited in the industry. However, it has been embraced by 3GPP Release 5 for media sessions authorization—that is, control of access to networks (possibly VPNs) used to offer multimedia services over third-generation systems.

## Captive Portal

The access control function can be enforced at the point of termination of access protocols by admitting network access only after the access protocol authentication phase has completed successfully. Otherwise, it is possible to enforce network access control via a *captive portal approach*. This method is commonly used in Wireless LAN-based access networks and broadband access networks. It forces users to authenticate by filling in a form on a Web page, which is the only place they can access after gaining IP connectivity. This can be done via a TCP redirect functionality on ports 80, 8000, and 8080, which are typically used for the Hypertext Transfer Protocol (HTTP)—the protocol used to transfer Web pages.

The TCP redirect functionality—implemented on a network device or a plain router placed at the edge of a VPN (or any IP network)—inspects all IP packets up to the transport layer. If the protocol is TCP and the port number is recognized to be HTTP, the network destination address and the HTTP header are modified to request the Web page that corresponds to the portal credential input page. Alternatively, when the network device is not built to operate at the application level, the portal itself may have the intelligence to always respond to incoming HTTP requests from not registered (admitted) IP addresses by sending via HTTP the sign-on Web page. Once user credentials are collected and the user is successfully authenticated, the network device, informed by the portal, lifts the TCP redirect functionality and lets user traffic flow freely, perhaps until some other event—such as user inactivity timer, usage time or volume limit overflow, an interval allowed before new authentication is required, or even redirection to some advertisement page—does not modify this state.

Without going into further details that may not be entirely relevant to the main topic of this chapter, we think that this way to perform access control will become more and more widespread because of the shift in access technologies toward broadcast media such as WLAN, and because sharing access network resources may require a provider selection phase before service begins. In addition, the user may be prompted at provider selection time with a second (personalized) Web page with information on his or her account profile, promotions, and links to partners offering other fancy services. This might be located on VPNs or service networks and not accessible until a new (captive) portal-based access control phase has not been successfully conducted.

Some service providers do not even use the complexity of captive portals based on TCP redirect, perhaps because the equipment they have bought only inspects packets up to the network layer and can be configured only to allow access to certain network layer destinations. Instead, they provide access to users if they deliberately point their browsers to a URL printed on a subscription card or a prepaid card they are given. Once they go through the authentication phase, the network allows the user identified by the source IP address (and perhaps also the MAC/Layer 2 identifier) that was successfully authenticated access to all destinations.

## Authentication

One of the most important functions VPN supports is authentication. In virtual private networking, every entity involved in communication must be able to positively identify itself to other involved parties and vice versa. Authentication is the process that allows communicating entities to verify such identities. One of the popular mechanisms for authentication widely used in today's networks is called PKI (see Chapter 2). This is known as *certificate-based authentication*, and the parties involved in communication authenticate each other by exchanging their certificates, which are guaranteed to be valid by a trust relationship with a certificate authority.

The authentication process may also involve providing shared-secret-based authentication information, such as a password or CHAP challenge/response pair, to an authenticator, such as an NAS, which in turn may look up a local file or query a RADIUS server. In this respect VPN operation encompasses two types of authentication: client-gateway and gateway-gateway. An example of client-gateway authentication in the GPRS packet data environment is RADIUS-based authentication of users accessing the GGSN. Only upon success are they admitted to an IPSec tunnel toward the customer network IPSec Gateway. The other case is common when site-to-site connectivity is set up, or when virtual dial-up networks are used and L2TP tunnel setup authentication is required between the L2TP Access Concentrator (LAC) and the L2TP Network Server (LNS).

## Security

From the previous discussion we know that VPN by definition is built across public shared unsecured facilities, which makes data integrity and encryption its mandatory requirement. The VPN can be secured by deploying one of the available encryption or ciphering mechanisms combined with secure key distribution systems. These mechanisms are covered in depth in Chapter 2. We must mention, however, that security is not limited to the act of encrypting VPN traffic. It also involves complex procedures by the operator and its suppliers (for instance, in delivering SIM cards with secret key verification and algorithms), and when the VPN is network based (see the section "Labeling (MPLS) and VPN" coming up in the chapter for a definition of network-based VPN), there must be a trust relationship set up between the service provider and the VPN customer that requires agreement and deployment of appropriate security mechanisms. For instance, the AAA server in the corporation may be accessible only by securing RADIUS messages via IPSec as they transit over the shared infrastructure. Also, the AAA server may belong to a network not included in the VPN itself, to allow for isolation of AAA traffic from user traffic.

## Tunneling as the VPN Foundation

Tunneling, originally covered in Chapter 2, is undoubtedly the single most important technology on which IP VPNs are built. Tunneling involves encapsulation of certain data packets within other packets according to a set of rules implemented at both ends of the tunnel. As a result, the contents of encapsulated packets become invisible to the public insecure network over which this packet is being transmitted. Examples of various tunneling technologies and definitions are provided in Chapter 2.

**NOTE** A tunnel can have multiple endpoints when a multicast destination address is used.

The tunneling concept as applied to virtual private networking is depicted in Figure 5.1. Here the packets sent from remote host A to host Z must traverse many other switches and routers. If router C encapsulates the packet coming from host A and gateway Y decapsulates it, the other nodes traversed by this packet will only recognize the "outer" encapsulating packet and will not be able to obtain any information about its payload or final point of destination. This way, the payload of the packets sent between C and Y will only be recognized by these two network nodes and the hosts A and Z representing the origination and destination points of the data traffic. This effectively creates a "tunnel" through which packets are transported with the desired level of security.

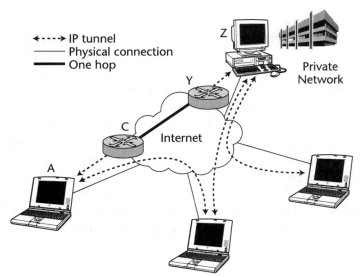

**Figure 5.1** Tunneling in virtual private networking.

The tunnel can be defined by its endpoints, the network entities where decapsulation occurs, and the encapsulation protocol being used. Tunneling techniques supporting VPNs such as L2TP or PPTP are used to encapsulate link layer frames (PPP). Similarly, tunneling techniques such as IP in IP and IPSec tunnel modes (all described in Chapter 2) are used to encapsulate network layer packets.

In the context of virtual private networking, tunneling may accomplish three major tasks:

- Encapsulation (just described)
- Private addressing transparency
- End-to-end data integrity and confidentiality protection

Private address transparency allows the use of private addresses over a publicly addressable IP infrastructure. Since the contents of tunneled packet and its other attributes—such as addresses—are understood only beyond the tunnel endpoints, private IP addressing can be completely masked from public IP networks using valid addresses (Figure 5.2).

Integrity and confidentiality functions ensure that an unauthorized party cannot alter the tunneled user packets during transmission without detection and that the contents of the packet remain protected from unauthorized access. Moreover, tunneling can optionally protect the integrity of the outer IP packet header, thus providing data origin authentication. For instance, in IP VPN, it is possible to use the IPSec AH header to protect the tunnel endpoints' IP addresses from spoofing. However, in the data industry, in many cases this has not been considered important, and actually many VPN gateways do not even implement AH (see Chapter 2 for more details on IPSec, ESP, and AH). The reason is that if the tunneled user packet is ESP protected, and the packet encrypted using secure key distribution and management techniques and algorithms that cannot easily be broken, such as 3DES, then any attempt to use IP address alteration to intercept or to send traffic would be purposeless. There would be no way the malicious endpoints could correctly participate in the IPSec ESP-based security association, and thus detection of the incumbent security hazard would be easy, and the certainty that it would be very difficult to interpret "stolen" user data would be very high. This is what matters to VPN customers and thus justifies the quite limited usage of AH.

**NOTE** AH may still be useful when tunnel setup control information needs to be provided. For example, this is mandatory to protect Mobile IP registration messages.

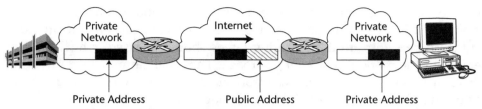

**Figure 5.2** Private IP address masking via tunneling.

When tunneling is applied to create a Mobile VPN, these four functions (encapsulation, private addressing transparency, end-to-end data integrity, and confidentiality protection) must be accompanied by a set of mechanisms to provide for dynamic tunnel switching or reestablishment to support VPN user mobility. Later in this chapter, we discuss the details of this additional requirement. From Chapter 4 you may recall that modern cellular packet data systems, such as GPRS or CDMA2000, already include such schemes based on GTP and Mobile IP, which were originally designed to support mobility. The mobile tunnels based on these technologies can be concatenated to static tunnels at the edge of wireless networks to provide a variety of MVPN architectures. In the following chapters we describe in detail how these mechanisms are applied.

### Labeling (MPLS) and VPN

The MPLS allows for nondestination IP address-based forwarding of IP packets over an IP backbone (see Chapter 2). One of the applications of this MPLS property is traffic engineering—that is, routing packets over paths determined by other criteria than mere optimal IP routing and perhaps based on the need to offer some level of QoS, or to select minimum-cost links or route over less utilized links to obtain load sharing. Another important application of MPLS is the provision of network-based VPN services between multiple-sites-based customer networks, also known as multisite VPNs. *Network-based* VPN, discussed in more detail later in the chapter, is a service offered by a service provider in an explicit way, via a provider edge (PE) router to customer networks. Provider edge routers normally connect to customer sites via customer edge (CE) routers via any Layer 2 or tunneling technologies, as well as MPLS.

The use of MPLS to offer VPN services is bound to the ability to control LSPs set up via some efficient VPN site's membership discovery/distribution protocol. There are mainly two schools of thought about the use of MPLS to offer VPNs. One camp thinks that a PE router fully controlled by the provider and equipped with multiple routing tables should be used, which we will refer to as the monolithic router approach. The other camp believes that the PE router should instead be based on virtual routers and that customers should have some degree of ability to manage them. We do not want to take a position on which is best, because, as it is often the case, practical considerations may suggest at times to use one method and at times the other. Rather, we will analyze pros and cons of each viewpoint.

The monolithic PE router approach is based on the ability of such router to support one routing table per VPN site, and each of these routing tables would advertise customer network site routes *intradomain* using the Internet Border Gateway Protocol (IBGP) and *interdomain* using BGP. In the case of a network prefix belonging to a site that overlaps with some other prefix of a different site advertised by the same PE router, the route would be accompanied by a route distinguisher to form the IPv4 VPN address family. This route may be associated to two additional BGP attributes—the *Target VPN* attribute and the *VPN of Origin* attribute—that would help in building a policy-based filter that can be used to configure VPNs on PEs. A label accompanying the route is also advertised, and the PE router includes its own IP address as the BGP next hop. The label is used as the inner label of a label stack that allows the IP packet to be sent from one PE to another PE, and to route the packet to the appropriate CE router based on the value of the label. This approach, described in [RFC2547], is widely implemented in the industry—largely because the leading industry vendor, Cisco Systems, is promoting it.

With the virtual router (VR) approach, the PE router supports software-based routers named virtual routers, each taking care of a particular customer network site. Since a VR is a regular router, customers can manage it and configure the forwarding equivalence classes (FECs) on it. Each PE VR belonging to a particular VPN can discover the other VRs through the use of OA&M, directory based methods, or a BGP approach such as the one described in [RFC2547].

For instance, it is possible to envision a GPRS operator who not only manages wireless access, but also makes its GGSN behave as the CE of a BGP MPLS, and uses MPLS VPNs to partition the network in VPNs devoted to different services. A GPRS operator may decide to create a MPLS VPN for intradomain packet data services that is offered to subscribers when they are not roaming and a VPN for roaming subscribers. This comes with the advantage that inter-PLMN backbone resources usage (such as IP addresses belonging to GRX blocks assigned to the provider for inter-PLMN operation) is minimized.

# Service Level Agreements

Entities involved in virtual private networking, such as wireless carriers, ISPs, corporations, and remote users, are bound by certain agreements to achieve the desired levels of service as well as the desired revenues for the services provided. Such agreements, drafted between all the interested parties and their partners, that define quantifiable and measurable levels of service are called service level agreements (SLAs). SLAs have been used in many forms for quite some time in networking. However, they are especially important in the context of virtual networks relying on a shared infrastructure or a multitude of shared infrastructures—as is often the case with MVPN. In mobile data networks where relationships between peers are already fairly involved, more than one SLA might be required to support all services and cover all entities that might be involved in such services on the provider or the consumer side. Let's take a more detailed look at the issues involved in setting up a comprehensive SLA in a mobile environment.

## MVPN SLA

MVPN SLAs are especially complex because they must include both wireless and wireline segments. Often, the performance of the wireless segment cannot be adequately enforced because of the inherently unpredictable nature of the air interface. Additionally, in the mobile environment a user may be roaming to a network outside the home mobile service provider's administrative domain, bringing up the issue of guaranteeing end-to-end service. The home service provider must somehow cascade the need to provide service guarantees in a framework roaming agreement, standard for all roaming partners. Such a roaming SLA would allow, with an acceptable degree of assurance, that the end-to-end service level guarantee can be met according to the performance figures promised to the customers. It is therefore likely that given the uncertain roaming and mobility patterns that may characterize mobile users and perhaps also because of unpredictable visited network conditions, Mobile VPN service providers may include in their SLA different service level guarantees for the cases when the user is roaming and when the user is in the home networks.

Following is a list of considerations you should take into account when compiling a SLA for a typical MVPN service:

- Fixed tunnel availability
- Fixed tunnel bandwidth guaranties
- Fixed tunnel latency

- Peak and sustained cell/packet rate
- Packet loss rate
- Session continuity guarantees (that is, a limit on the expected times the session could be lost within some coverage areas and under some constrained wide area mobility conditions)
- Idle sessions timeouts (which may differ from those normally enforced by the corporate network access server, because of the need to save resources on the wireless network side)
- Allowed roaming regions and performance when roaming

Having analyzed main VPN properties and underlying technologies, let's turn our attention to VPN classification.

# Classifying VPN Technology

There are two approaches to the classification of VPN technology:

- *Architecture taxonomy* deals with how VPN is architected and deployed.
- *Tunneling taxonomy* deals with how underlying tunneling techniques are implemented.

Historically, architecture taxonomy is more often used in sources dealing with wireline data VPNs, while tunneling-based taxonomy is usually applied in sources addressing cellular systems. While this book mostly deals with mobile wireless systems, we will provide an overview of both taxonomies and their applicability to Mobile VPNs. The examples of VPN classifications include compulsory versus voluntary, which is based on tunneling taxonomy, and site-to-site versus remote access, which is based on architecture taxonomy.

## Tunneling Taxonomy

All IP VPNs can be implemented using basic tunneling methods:

- End-to-end, or voluntary
- Network-based, or compulsory
- Chained or mediated tunnels, which fall somewhere in the middle

Based on the tunneling method being used, the VPN itself can be classified as voluntary, compulsory, or combined. We will now take a closer look at all three methods.

## *Voluntary VPN*

*Voluntary IP VPN* provides remote users with the ability to create a tunnel from their terminals, such as mobile phones or PDAs, to certain tunnel termination point, such as a VPN gateway that resides within the private network. Private networks are usually protected by firewalls and require firewall traversal and security mechanisms—for instance, user authentication and data integrity and confidentiality protection—applied to the remote access traffic. Consequently, remote user equipment must support proper protocols to satisfy these requirements. For example, a remote user equipped with a device such as a PDA could establish an IPSec ESP tunnel to a corporate network using PKI-based key distribution (also known as an asymmetric keys approach) or predistributed shared secret key (also known as a symmetric keys approach). All data between such mobile stations and the private network would then be encapsulated in the secure end-to-end IPSec tunnel. The end-to-end tunneling in this example exists only for the duration of the session and is torn down when the remote users do not require private network access, or when the user must be preemptively disconnected based on a set of predefined events, such as session duration or limits in access rights.

This type of VPN service is depicted in Figure 5.3, which uses mobile dial-up access over a GSM network as an example. In this scenario, the remote user establishes a VPN connection to a private network after a wireless carrier grants him or her Internet access. Note that both wireless and wireline-based access to the Internet allow roaming users to establish this type of VPN at will, by "voluntarily" opening a communication channel to the private network when they need it—hence the name of the approach.

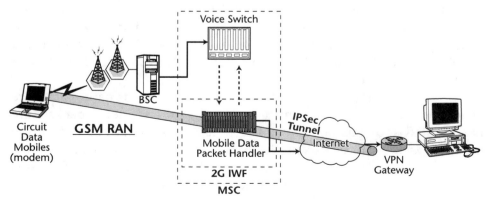

**Figure 5.3**   Voluntary VPN over 2G network.

Voluntary VPN carries a number of significant advantages. For private network IT administrators and often for remote users, this is the simplest way to establish a remote access VPN. Remote users simply need access to the Internet or any other public IP network, and a VPN client in their mobile or fixed devices. All that the private network IT department needs to do is to provision a VPN gateway connected to the Internet and capable of terminating a particular type of tunneling, and establish a proper set of policies and security procedures. The service provider offering Internet access service cannot access the end-to-end encrypted private data being transmitted between remote user and private network, and hence it will not have to be entrusted with it. Voluntary VPN also does not require any preestablished relationship between corporations and service providers. Therefore, there are no multiple SLAs and legal agreements about data confidentiality. However, the user and the corporation should be ready to accept a network access service that may be less predictable and often qualitatively inferior to the service provided to parties that instead enter in a SLA, unless the service provider offers predefined levels of service such as a "business class" Internet access option.

Voluntary VPNs require that public, topologically correct IP addresses be assigned to remote users' equipment. This requirement—along with other properties of an end-to-end tunneling—creates a number of potential drawbacks to voluntary VPN service. Because of the limited number of IPv4 addresses available to service providers—especially for mobile operators that would like to offer "always on" connectivity to the Internet to their subscribers—reliance on private addressing schemes to conserve valuable IP address space, combined with various subnetting and address translation techniques such as NAT, is very common. Until recently, this made end-to-end tunneling, where public IP addresses were required, impossible. For example, IPSec AH mode is not compatible with NAT, which would not even allow for the tunnel to be established. Luckily, a few mechanisms allowing for NAT traversal have been recently introduced by the IETF and are being widely implemented by the industry (see Chapter 2). Also, the emergence of an IPv6-based network and the expected gradual conversion to an all-IPv6 Internet should resolve problems with insufficient addresses.

Another disadvantage of voluntary VPN arises from the nature of secure end-to-end tunneling, in which the contents of the tunneled packets are encapsulated and thus not available for inspection by any nodes on the tunneled packet's path except the tunnel endpoints. This makes QoS, classe of service (CoS), and the majority of traffic-shaping mechanisms requiring multifield packet inspection a difficult to impossible task.

Monitoring equipment and certain firewalling functions will fail to work properly as well. We discussed some QoS issues that relate to tunneling in Chapter 2.

When Mobile VPNs are implemented in a cellular wireless environment, voluntary tunneling will lead to an extra layer of encapsulation over the last-hop wireless link. This will consume more of already scarce and expensive radio resources. Also, complex encryption and security algorithms may not be suitable for implementation in small wireless devices, which typically have limited processing and battery power. Additionally, widely mutable radio conditions and lossy wireless environment are not friendly to the establishment and preservation of IPSec tunnels. This may translate in a long tunnel setup time, or in extreme cases, to complete failure and perhaps the need to move to a region with better coverage. Note that this is not only affecting cellular systems but also Wireless LAN-based access networks.

For these reasons, while voluntary tunneling provides a clean and secure end-to-end solution for access to private networks, often greater VPN efficiency and unique services can be achieved with a participation of service providers prompting an introduction of another VPN type.

## Compulsory VPN

A service provider may offer compulsory VPN service by concatenating or chaining multiple tunnels or provisioning a single tunnel for a *part* of a data path between two participating endpoints. For example, a compulsory VPN can be based on a tunnel created between a private network and a service provider and not extended to reach all the way to a remote user that is using the network access service. As a result, with *compulsory VPN service* the remote user does not need to have any involvement into VPN establishment process and is "forced" to use the available preprovisioned service whenever the access to the private network is required, hence the name.

This VPN type assumes that the operator's network infrastructure features the intelligence and functionality necessary to support VPN services based on the tunnels or sets of tunnels provisioned between the private network and service provider's networks rather than all the way to the end-user device. In both cases, the enterprise must preestablish a detailed SLA with the service provider responsible for VPN service and must trust it to handle its valuable data with the necessary care and confidentiality. The service provider often participates in the network's access control, and the corporation must trust the service provider to deny access to nonauthorized users according to the network access policy defined by the corporate network administrator. One possible compulsory VPN scenario implemented in the CDMA2000 infrastructure—based on Mobile IP—is depicted in Figure 5.4. In this figure the user data is encapsulated into Mobile IP tunnel only between the PDSN in carrier's network and the HA owned by corporation.

The need to keep a part of the private data path unprotected, to trust the service provider, and to establish multiple SLAs and complex data confidentiality agreements are some of the drawbacks of compulsory VPN. In mobile environment, security problems become even more serious, since the user traffic is being sent over potentially insecure radio channels. During packet data roaming, the unprotected traffic to and from the mobile station must also traverse the visited carrier network (which may or may not have established SLA with the corporation served by a home wireless carrier) before being tunneled to original carrier's network. If there are insecure links in this network, especially unencrypted links in the backhaul section, this could present serious security problems. These problems are hard to address unless the provider puts careful service provisioning in place. For instance, gateway-to-gateway IPSec tunnels in critical points of the network could address this problem.

On the positive side, the compulsory approach better utilizes the air interface by avoiding over-the-air encapsulation overhead, which is especially advantageous for cellular wireless systems, and by simplifying the user equipment. When compulsory VPN is used, the end-user equipment does not have to support any VPN clients or tunneling or security capability at times they could be CPU-hungry and battery-life-consuming. Also, the user is not involved in VPN creation and only needs to request the service when accessing the service provider's network.

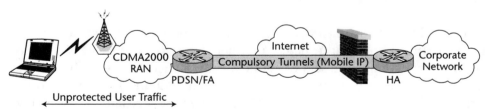

**Figure 5.4** Compulsory VPN in CDMA2000.

Compulsory VPN presents a number of other significant advantages to service providers. Offering and marketing compulsory VPN as a feature can potentially enable new business models and carrier service offerings. With the voluntary approach, service providers do not get involved in provisioning and often are not even aware of the existence of encrypted and encapsulated traffic unless they offer special access points to the Internet associated to publicly routable IP addresses or NAT traversal-compliant devices. In contrast, compulsory VPN access offerings can be marketed in different forms by carriers to a variety of private enterprises and ISPs interested in outsourcing their remote access function. This will bring new revenue streams, along with greater differentiation from the competition service offerings.

Another benefit of compulsory VPN for service providers lies in greater control over the user. In a compulsory model, the service provider is usually involved in user authentication and IP address assignment (though the latter might be a mixed blessing in some situations), which allows it to control user provisioning to a greater extent. IP addresses can be assigned to remote users from the customers' networks private address space, thus saving the usage of publicly routable IP addresses from the provider side.

### Chained Tunnel VPN

The third type of VPN is not easily classified as either voluntary or compulsory. *Chained tunnel VPN* consists of a set of concatenated tunnels that extend all the way to the end-user equipment. Chained tunnel VPN can come in many different forms, as shown in Figure 5.5, which depicts a few tunnel chaining options in the GPRS network.

Similar to the voluntary VPN approach, chained tunnel VPN provides end-to-end user data protection, and the user participates in tunnel initiation. Like compulsory VPN, the service provider is involved in chained tunnel VPN provisioning and construction and can easily apply QoS and traffic shaping at the tunnel concatenation points. This participation does necessitate SLA and data handling agreements, though.

In our opinion, all of these VPN types have their positives and negatives and will coexist in future networks. The service providers can offer them interchangeably, depending on the available technology, suitability to task, and business environment.

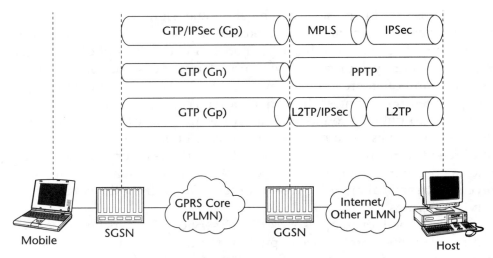

**Figure 5.5**   Chained tunnel VPN options (GPRS environment).

## Architecture Taxonomy: Site-to-Site and Remote Access VPN

In this section we look at VPN classification from an architecture rather than a tunneling perspective. We also analyze the current wireline IP VPN taxonomy and its applicability to the mobile environment. VPN architectures can be classified into two main types: Site-to-site VPNs, also called LAN-to-LAN or POP-to-POP, and Remote Access VPNs. *Site-to-site VPN* can include variations usually referred to as *Extranet VPN* and *Intranet VPN*, which share the same properties but are designed to solve different sets of problems. *Remote Access VPN* includes dial-up and direct packet access methods, which are also discussed in a main architecture framework. All of these VPN types can be implemented (at least in theory) over mobile wireless networks; however, the discussions involving MVPNs are concentrated mainly around remote access type, since normally a wireless network connects a single remote host to a network it wants to access.

### Site-to-Site VPN

*Site-to-site VPN* is used to connect geographically distributed corporate sites, each with private network addresses administered in such a way that normally conflicts do not arise. In traditional networking, remote corporate offices can be interconnected by partially full meshed networks based on T1 and E1 leased lines or link layer circuits, such as ATM or Frame Relay PVCs inside the service provider's networking "cloud." Another option is to implement it based on private routed network or based on MPLS label switched paths, for instance, à la [RFC2547]. Private routed networks require routers capable of segregating traffic from different intranets, based on multiple routing tables, leading to scalability problems. The resulting infrastructure is called an intranet.

**NOTE** Because of the recent slew of company mergers, acquisitions, office up- and downsizing and the shrinking pool of available IP addresses, a NAT in the future may be more necessary than ever to exchange traffic between intranet or extranet sites.

Similar results can be achieved by provisioning secure IP VPN tunnels over public Internet or service provider IP networks. Here, IP VPN provides significant cost advantages along with greater simplicity. Communication costs are reduced because the customer pays only for the access to the service provider's network IP network or Internet access. Remote offices are connected via tunnels provisioned over a public shared IP network, such as the Internet or a commercial shared IP network offered by service providers like WorldCom or AT&T. Along with public IP network access, VPN implementation requires CPE, in the form of VPN gateways capable of supporting sufficient numbers of individual tunneling sessions at each site.

Site-to-site VPNs are still possible in mobile environments, but the value of such models from a wireless subscriber perspective is still unclear at best and they are outside the scope of this book. It is nevertheless possible for this VPN type to be deployed in the context of wireless core networks such as GPRS PLMN and GRX, when a wireless service provider wants to deploy alternative connectivity to its roaming partners. In addition, a GPRS access point to a corporate network may be considered as a particular instance of a site in a site-to-site VPN. In fact, a service provider may offer full connectivity to all corporate sites and even route packets directly to the most appropriate customer site via IP tunnels that connect them to an access point.

### Extranet VPN

*Extranet* is a relatively recent concept in data networking. It is usually deployed in the situation when a corporation needs to interact not only with its own remote offices but also with the sites, which belong to its customers, suppliers, and other entities it is engaging in transactions or information exchanges with. These are generally referred to as *partner networks*. To support such communications, VPN tunnels can be established between private networks that belong to different private entities. VPN functions such as access control, authentication, and security services can be used to deny or grant access to resources required to conduct business. Security threats to the extranet—including unauthorized resource access—are greater than in an intranet, so the VPN and extranet environment should be carefully designed with multitiered access control policies and unique security arrangements between extranet members. For instance, a supplier may have access to the customer's order and perhaps inventory system, while the customer may want to be able to track the supplied material by accessing the suppliers' delivery status tracking system.

IP VPN's ability to provision tunnels and access control preferences dynamically within minutes or even seconds and often without the need to notify the access provider (if the permissions to do so were part of the SLA in the case of compulsory VPN) is especially useful in the extranet environment. Dynamic provisioning is crucial for today's business communication where business needs and situations can change quickly, requiring new networking arrangements to be reestablished literally on the fly. This capability addresses networking technologies disadvantages, including relatively long provisioning intervals and complex provisioning procedures. For example, a Frame Relay PVC provisioning typically requires somewhere between 5 and 20 business days to complete. Figure 5.6 shows a major corporation establishing dynamic relationships with its suppliers and other business partners.

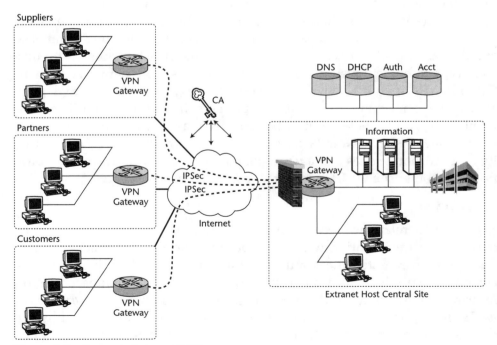

**Figure 5.6**  Dynamic Extranet VPN.

Note that the use of IP VPNs interconnecting partner networks is worthwhile when a sizable amount of information needs to be exchanged between networks. When the interaction between partners can be performed via Web-based applications, the preferred approach is normally to set up TLS- or SSL-protected Web interfaces to applications and information repositories. It is also common today to base business interactions on customers' or suppliers' portals.

### Intranet VPN

A corporation may decide to set up a VPN not only to interconnect sites belonging to the corporation but also to define virtual networks within its own administrative domain. Most of the time this is not required, since security zones may be defined by means of properly provisioned firewalling and resource access control policies at the application level. However, in mission-critical environments, it is possible that organizations—especially large ones—may decide to enforce security measures by segregating particularly important traffic and network zones via IPSec-encrypted tunnels and enforce on these network zones additional authentication and access control policies. Companies need to ensure the confidentiality of such data as employee personal records, as well as information about the upcoming events such as product launches and reorganizations. Intranet VPN's main role in the corporate network is to establish and manage different levels of internal access to specific information. In this capacity, the Intranet VPN is yet again used to create an environment similar to physically segmenting groups of users on distinct LAN subnets joined by bridges and routers.

## Remote Access VPN

Remote access networking was originally conceived as a way to provide remote hosts with the access to the information resources and data services located in a private network. Remote access is equally useful for both consumers and remote workers or other business and institutional users. Corporations and other entities had to maintain adequate facilities, usually consisting of farms of remote access servers (RASs) and appropriate security equipment, to maintain the proper level of service availability and reliability to remote users. Dial-up connections were considered private and adequate for all intents and purposes (recall our opinion on popular misconceptions about private networks). Wireless dial-up technology, explained in the previous chapters, worked—and still does—very similarly to wireline and presented similar convenience, as well as a similar set of problems.

### Dial-up Remote Access VPN

Dial-up remote access is expensive and requires significant support from corporate IT departments. Often, remote users are far away from corporate headquarters or data centers and require long-distance or 800 calling, which can become especially costly for international callers and telecommuters who tend to stay connected for a long time. Corporations relying on this technology were frequently forced to create (or outsource) multiple regional data centers and maintain local support and maintenance personnel to avoid high long-distance charges. Dial-up access also requires expensive RAS equipment, which is complex and far from being trouble-free.

With the rapid buildup of remote access networks—complete with nation-wide and even international rollouts of local dial-up points-of-presence (POPs) by service providers—this situation has changed dramatically. Now all the worries about the dial-up procedures could be offloaded into the capable hands of ISPs and other network access providers for whom dial-up or for that matter any type of Internet access was not a burden but a core competency and who already had extensive regional dial-up facilities available. Remote access outsourcing was born, and then it was reincarnated a few years later as Remote Access IP VPN.

Dial-up remote access VPNs can be based on either compulsory or voluntary tunneling methods. In a typical dial-up access outsourcing (compulsory) scenario, the user dials into a local Internet service provider's POP, establishing a PPP link. After the user is authenticated and a PPP link is established, the service provider establishes in a compulsory manner—that is, transparently to the user—a tunnel to a gateway in the private network, the remote user is wishing to gain access to. The private network performs final user authentication and establishes connection. The resulting architecture is depicted in Figure 5.7. In contrast, traditional direct dial-up access involves a user dialing into a bank of modems, a RAS, or a concentrator located within a corporate data center. The tunneling technology of choice for dial-up access outsourcing VPN is L2TP, which is widely accepted by both wireline service providers and corporations. This technology is described in detail in Chapter 2. To offer this VPN option to remote workers, corporations must establish detailed SLAs (for instance, to configure a list of LNSs, security aspects, and QoS levels), with one or more service providers, which will be responsible for local dial-up termination facilities.

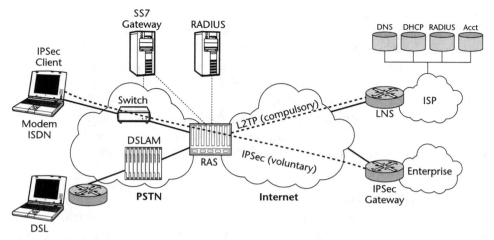

**Figure 5.7**   Landline remote access outsourcing.

The approach resembling wireline dial-up remote access was adopted for 1-G and 2-G circuit data systems to offer Remote Access VPN services to mobile users. IWFs based on familiar RASs were outfitted with tunneling hardware and software, which allowed wireless carriers to avoid costly and cumbersome IWF dial-out procedures (see Chapter 4). After the operator establishes proper SLA with a corporation, the mobile user's data traffic could be compulsory-tunneled to their private networks using L2TP technology in the way already familiar to corporate IT departments experienced in wireline remote access. The operators who had followed the model now had the option of aggregating the user data traffic onto their backbone network, which would allow them to further optimize their data backhauling costs. A good example of such architecture is CommWorks Quick Net Connect based on Total Control IWF, mentioned in Chapter 4.

Voluntary VPN is essentially access technology independent, and it can therefore be easily supported over any dial-up connection, including wireless. All end users have to do is establish a dial-up connection to the ISP of their choice and then via a VPN client establish a secure network layer tunnel to a particular private network. There are ISPs with a global footprint that offer this kind of capability in wireline data networking. Otherwise, consortia of ISPs—like those that are part of iPass or GRIC—may offer global roaming capability. While widespread in the landline networking, this technology has been of limited applicability in the cellular wireless environment. The reason is in the limited data throughput available to the *circuit* wireless data users. Voluntary VPN relying on an end-to-end tunneling requires extra encapsulation on a last-hop wireless link, further decreasing already-low throughput and making this option suitable for vertical applications but not for a mass deployment. We expect this trend to change, however, with the wider deployment of higher-speed wireless data systems.

### Direct Packet Access VPN

During the past decade, new Internet access technologies such as Integrated Services Digital Network (ISDN), Digital Subscriber Line (DSL), and cable became available to telecommuters and other remote workers who traditionally relied on dial-up access. Both service providers and corporations had a remote access technology option that allowed more efficient use of both compulsory and voluntary VPN for secure private network access. Wireless packet data systems by design are closer to high-speed wireline services such as DSL and cable modem and offer mobile users the same level of convenience, such as automatic always-on capabilities, granular billing options, and banishing of frequent logon procedures. Most of the discussion in the following chapters is devoted to this type of remote access provided within the wireless packet data systems framework.

## Moving from Wireline to Wireless and Mobile

We are now ready to apply the principles of VPN technology to the mobile wireless environment. In the previous section, we addressed main properties and design methodologies of VPN provided in various dial-up environments—including general circuit-switched wireless data systems. The rest of this chapter (and the book) focuses on VPN support over cellular wireless packet data networks such as those incorporated into 2.5G and 3G systems. We start the discussion with a terminology clarification, followed by the introduction of basic VPN mechanisms into main wireless packet-data-based systems. This should provide you with enough background and a good transition to the detailed discussions in Chapters 6 and 7.

### Wireless versus Mobile

Let's first clear up confusion between the terms *wireless* and *mobile*, which are often inaccurately used to refer to specific network types. Many publications use the terms interchangeably. This is valid in certain areas and circumstances. However, it is our intention to use these two terms in a way as precise as possible. When applied to data networking, wireless and mobile mean two different things. Wireless networks can be mobile or nonmobile; for instance, microwave, Local Multipoint Distribution Services (LMDS), line-of-sight laser communication, or Bluetooth links are typically not supporting mobility and are simply used to provide fixed network access in the absence of wire or fiber. Mobile networks can be established over wireless as well as wired infrastructures as long as user mobility or network access service roaming are supported.

The term *wireless* means exactly what it says: something that can exist without wires. So, for example, *wireless network* should refer to a network built over the air (or in space in case of satellite communications). The term *mobile* refers to something that changes its location over time, so *mobile host*, for example, refers to a host whose position in the network changes over time relative to other hosts. *Data mobility* in turn can be defined as the ability of a host to change its point of attachment as well as its method of communication to the network while maintaining uninterrupted connectivity (recall the definition in Chapter 1). Systems capable of mobility support are therefore called *mobile systems*, and the end-user devices supporting mobility are called *mobiles*, *mobile stations*, and *mobile nodes*.

**NOTE** Now as the definitions are established we should mention that in real life, mobile networks are almost always implemented in the wireless environments, while wireless networks can be both mobile and fixed.

As is often the case, these definitions can be widely interpreted and may produce more questions than answers, including:

- Would a change of IP address constitute a service interruption?
- Can mobile data technology that spans only one wireless system be considered truly mobile?
- Can mobility be supported and what would it be used for over fixed networks such as corporate LANs?
- Is it important to be able to contact a user asynchronously anytime independently of his or her current location?
- Is mobility to be supported on physical, link, or network layers?

The rest of this book could be devoted to the discussion of this fascinating topic, but we need to continue on with the topic at hand. We will try to resolve as many questions arising from these definitions as we possibly can, time and space permitting.

## Significance of VPN in the Wireless Packet Data Environment

Recall from Chapter 4 that cellular packet data technology is based on a dynamic tunneling concept in which sequential tunnels are created and torn down between the foreign networks visited by the mobile station and its home network. The complexity of the task of providing VPN service in such environment lies in somehow combining this technique with fixed or "quasi-fixed" tunneling topology required on the wireline side to provide mobile users with secure private network access. This task becomes especially complex when compulsory VPN service is required. In this case wireless operator must possess the equipment capable of supporting not only *dynamic tunneling* but also *dynamic tunnel switching* between dynamic and fixed portions of their infrastructure. Figure 5.8 provides a sample architecture illustrating this requirement. In the case of voluntary VPN, this task is simplified a bit, since end-to-end tunneling is generally unaware of the underlying infrastructures as long as the endpoints' IP address stays fixed. This requirement should be addressed by data mobility schemes such as those used in GPRS or CDMA2000.

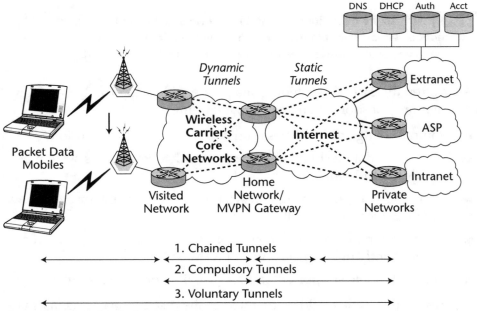

**Figure 5.8**  VPN in wireless packet data environments.

Let's now analyze why MVPN in packet data systems is necessary. Supporting MVPN requires complex tunnel-switching-capable network nodes and mobile devices. In wireless packet data, the mobile stations are enabled by the network layer mechanisms to change their location and point of attachment to the network while maintaining the connection to their home network. Often the mobile station roams onto foreign networks supported by operators other than its original carrier. While roaming, the MS keeps its connectivity to its home network through the use of tunneling schemes supporting mobility such as GPRS in GSM and UMTS systems or Mobile IP in CDMA2000 system. In such an environment, it is impractical to establish any kind of permanent dial-up-style circuit connection between the MS and private network. That would actually defeat the purpose of migration from circuit- to packet-switched environment.

The best—and arguably the only—technology for private network access in such an environment is MVPN, either compulsory or voluntary, based on mobility-enabled tunneling protocols at least over some leg of the end-to-end data path. Thus, MVPN in packet data systems would not just be an access option—as it is in wireline networking, along with other types of access such as direct dial-up, leased lines, ATM, and Frame Relay—but a necessity. Having summarized the importance of MVPNs, we can now take a more detailed view at main MVPN types.

## Voluntary MVPN

MVPN based on voluntary tunneling is implemented almost exactly the same as wireline VPN. As with wireline VPN, it's important to consider whether the service provider uses public or private IP addressing schemes and what IP address translation mechanisms (if necessary) are used when private addressing schemes are in place. Another consideration, which is unique to the wireless environment, is the stability of the IP address assigned to a mobile device. Generally, tunneling protocols supporting mobility in modern packet data systems help to keep IP addresses assigned to mobile stations constant. Some even allow for permanent fixed IP addresses to be preprovisioned in the end user devices—a perfect condition for stable end-to-end tunnels forming a foundation for reliable voluntary VPN.

However, in some wireless systems certain access modes provide only limited IP mobility. Remote users and IT managers must take this into consideration when ordering wireless service with the intention to use it for voluntary VPN. For example, in CDMA2000, *Simple IP* access mode provides user mobility only within boundaries of the same PDSN/FA (more on this in Chapter 7). Here end-to-end tunnels cannot be kept alive when changing the serving PDSN, since the packet data bearer needs to be reestablished toward the new PDSN and the MS must acquire a new IP address. This also requires the VPN client on the MS to restart the session with the new IP address. That might not appear to be a significant problem given that a typical PDSN usually may span areas of several hundred square kilometers, but for mobile users traveling along the boundaries between two different PDSNs, this might present a real challenge. Using a compulsory VPN technology with a Simple IP access mode would not ameliorate the situation either, since the end-to-end connectivity is lost and a new remote network access procedure needs to be restarted each time a new PDSN is visited. This can be changed if a Mobile IP-based access mode is used. Mobile IP access mode can be used two different ways. One provides direct access to the network the end user needs to access, while providing mobility. The other way is plain but uninterrupted access to the Internet offered by the mobile operator, in which the user can choose to voluntarily set up an end-to-end tunnel using a common VPN client.

Another interesting aspect of voluntary MVPN access to networks over wireless links is its moneymaking properties from an operator perspective. It is somewhat likely that consumer access to the Internet will be differentiated from the Internet access service that allows a customer to use a VPN client, since the potential revenues per subscriber from consumer access are likely not to be the same as those from business customers accessing their corporate network using a VPN client. For this reason, wireless carriers need to offer distinct access points to the Internet for these two classes of users. The network access attributes will likely be different at these access points, such as public/private IP address, NAT with or without the NAT traversal feature, firewalling, and virus filtering. From an operational point of view, this may be enforced at contract signup time via the appropriate definition of the subscribed services.

## Compulsory MVPN

As mentioned at the beginning of this section, compulsory MVPN is based on the same principles as its wireline counterpart. However, while wireline compulsory VPN is based on a single fixed tunnel (or more rarely on a set of fixed concatenated tunnels), the MVPN implemented in a wireless environment is different and is based on a combination of dynamic tunnels supporting mobility and fixed tunnels on the wireline side. To achieve this functionality, referred to as *dynamic tunneling switching*, wireless operators must deploy intelligent infrastructure elements capable of supporting a variety of tunneling techniques.

Tunnel switching is a fairly recent concept, initially introduced by wireline data communication vendors competing in the IP services market. Devices supporting tunnel switching capability must be able to route user data through sets of tunnels by terminating the tunnels carrying the incoming data and originating the tunnels encapsulating the outgoing data. This is usually based on a set of policies provisioned in the network or individual devices by the wireless carrier on the business customer's behalf. Alternatively, instead of using tunnels, a private network of leased physical lines or ATM or Frame Relay VCs between the wireless operator's access gateway and the enterprise could be deployed. However, the leased line approach would eliminate VPN cost advantage and ease of management and provisioning, which is especially important in the mobile environment.

Compulsory Mobile VPN may be implemented differently depending on the allowed mobility pattern. For instance, it is possible to build a compulsory service in CDMA2000 Simple IP access mode given that user mobility would be limited. This is often the case for business users accessing corporate networks from hot spots like airport lounges or hotel rooms. This service requires dynamic establishment of an L2TP tunnel between the serving PDSN and the customer network (see Chapter 7 for details on how this is done). In fact, it is not possible to allocate a particular PDSN where a static compulsory tunnel between the corporation and the wireless carrier can be defined, since the subscriber may be using any PDSN serving the radio access network where he or she is located.

Architecturally the compulsory service may be implemented in CDMA2000 or GPRS systems by enabling the device that is a termination point of the protocols supporting mobility, such as PDSN or HA and GGSN, respectively, to be also the originating point of traffic exchange with customer networks via a set of fixed tunnels.

The compulsory model does not fit well with the currently commercially deployed WLAN access mode, since current access control methods require the user to first acquire an IP address and then authenticate at a portal to gain network access. This requires the use of a DHCP-assigned IP address that belongs not to the corporation, but to the network that the subscriber is visiting. Therefore, it will not be possible to admit the user to the corporate network. This would create many security problems, since source-address-based firewalling and packet-filtering rules would be no longer possible, making this unacceptable for corporate IT departments. One way out of this situation is to use Mobile IP-based mobility, since Mobile IP comes with its own access control and the MS can be assigned an address belonging to the corporate access network. While it is hard to judge which of the myriad MVPN options will become the standard, the final outcome will be—as always—shaped by the market and the needs of real-life customers.

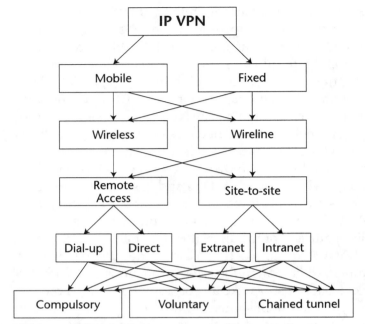

**Figure 5.9**   VPN family tree.

## Summary

We would like to end this chapter with a diagram depicting the IP VPN family tree (Figure 5.9). The goal of the discussion in this chapter was to help the reader better understand general VPN technology, taxonomy, and terminology and then to add mobility into the mix by introducing Mobile VPN. Here we provided an overview of the VPN enablers, such as tunneling and security, as well as main VPN topology examples, such as site-to-site extranets and remote access outsourcing. Figure 5.9 summarizes most of the discussion we had and creates a clear VPN hierarchy, which should serve as a good foundation for the material presented in the following chapters, which discuss Mobile VPN in much greater detail.

# GSM/GPRS and UMTS VPN Solutions

Expected subscriptions for Global System for Mobile communications (GSM)/General Packet Radio Service (GPRS) and Universal Mobile Telephony Service (UMTS) systems spurred unprecedented capital investment in the UMTS spectrum. A number of European Union governments have enjoyed a staggering surge in income by selling UMTS spectrum usage licenses (see Table 6.1 for the UK license bid final results), resulting in significant debt and a subsequent high-risk that cellular system operators now have to deal with. As a result, the entire industry is fast becoming short-term positive cash flow driven industry. This compels operators to:

- Open the network to application and service delivery partners
- Deliver *business quality* and personalized network access services to corporate users

GSM/GPRS and future UMTS operators need to deploy the most sophisticated and flexible IP services architectures available in the industry in order to have enough ammunition to fight a fierce battle to win customer base, to keep churn low by providing high customer satisfaction and service access predictability, and to sustain high profit margins. In this

**Table 6.1**   Final Bids on UK Spectrum Licenses

| LICENSE | UK 3G LICENSE WINNERS |
|---------|----------------------|
| License A | Canadian group TIW, secured A for £4.38bn. Bidding for this license, which offers the best range across the radio spectrum, was reserved for a new entrant into the UK market. (License currently held by Hutchison 3G Ltd.) |
| License B | Vodafone won the battle for B, which offers the highest bandwidth, with a £5.96bn bid. The UK mobile service provider beat off strong competition from BT. |
| License C | BT3G settled for a lower-capacity license with a £4.03bn bid. |
| License D | One 2 One, a UK mobile operator, made a £4.003bn bid. |
| License E | UK mobile operator Orange made a £4.09bn bid. |

chapter we provide examples of how these architectures can be designed to support new ways to keep revenues high in such a competitive service delivery landscape. We also describe methods to provide advanced data services, such as virtual private networking using both circuit- and packet-switched data access provided by cellular networks. A case study is included at the end to give you a working example of the architecture.

# GSM and UMTS Circuit-Switched Data Solutions

Circuit-switched data (CSD) has been the norm in cellular networks for a few years now. In GSM, 9.6 Kbps and 14.4 Kbps are the available CSD bearer services. These channel bit rates do not supply adequate support for today's bandwidth-hungry applications. They were, however, considered sufficient at the time when the GSM CSD bearer service was defined, since at that time the speed of wireline modems was similar. UMTS defines larger-capacity CSD bearers of 64 Kbps and above. Both GSM and UMTS packet services oriented bearers—supported by GPRS and the UMTS PS domain—can be used to provide data services and foundation for Mobile VPNs. Today CSD is mostly used for WAP-based applications.

In this section we briefly describe CSD-based VPN technology. Deployment of CSD-based VPNs will not be mainstream in the future; however, it is interesting to note the integration of the legacy into the new VPN paradigms.

## CSD Solutions Technologies

In Chapter 4 we described the technologies related to CSD bearer channels. To provide access to a data network, the CSD bearer on the wireless access network side needs to be terminated at an Interworking Function (IWF), where direct access to the network or conversion to wireline CS protocols takes place. Typically, PPP is used to provide link layer and host authentication configuration services over CSD bearers. In such environments there are two approaches to providing VPN services:

- The IWF could terminate the wireless bearer, convert it to an ISDN circuit (for instance, part of a PRI link to some switch), and then terminate ISDN calls to some RAS device that can in turn act as an LAC or simply provide network access services.

- The IWF itself could provide LAC functionality and handle L2TP tunnels established toward LNSs in customer networks or within the operator network.

The latter approach proves more effective as IP-based VPNs are becoming more and more common in the marketplace and corporations and service providers are streamlining the operation of their networks to reduce the number of different technologies that needs to be handled. It is also an effective way to provide WAP service, where the WAP gateway tightly interacts with the LNS (perhaps via the AAA infrastructure), in order to enable user identity information usage. The commercial success of these IP VPN solutions for CSD has been quite limited, but it is still worth mentioning, since this method proves to be very well suited for integration with equivalent services in the *packet-switched* technologies.

## CSD Deployment Scenarios

Two areas of CSD-based VPN usage are possible: corporate network access and WAP-based services provisioning. WAP service provisioning can also use Unstructured Supplementary Services Data (USSD), and even SMS, but these channels are more appropriate for very low bit-rate applications, transactional applications, or to support a *push channel* to the terminal. Web browsing mediated by WAP gateways normally requires CSD. However, the WAP protocol is also used in GPRS deployments, and WAP-based applications are commonly used in current GPRS terminals equipped with basic user interfaces and displays. It is expected that WAP, despite the consensus that it is not to be the ultimate delivery vehicle for mobile data applications, will stay alive for a while, at least until an adequate replacement is found that can provide similar functionality in a standard way.

The need for CSD VPNs typically arises when the network used to provide CSD-based access is shared for other applications. For instance, the network between the IWF and the WAP gateway may be used for other purposes than simply carrying WAP traffic, and the same IWF could support connectivity to multiple WAP gateways, possibly outside the domain of the network operator itself. For instance, some banks may require terminating the circuit—possibly over an L2TP tunnel—at a WAP gateway or a RAS owned by the bank itself. The same network could also be used to tunnel traffic back to corporate data centers (possibly hosted in outsourcing arrangements by some ISP), third-party ASP networks, and ISPs. A different telephone number is normally required to access different services when a CSD bearer is used. The telephone number may be associated to an NAS, an LAC, or a WAP gateway access point.

Figure 6.1 illustrates the case where the CSD traffic is tunneled via L2TP to an LNS that provides access to a WAP and value-added services network. The same network can be accessed using L2TP from a Gateway GPRS Support Node (GGSN), that is, via packet-based access. Thus, it is possible to define an L2TP-based IP VPN that allows the provisioning of the same services.

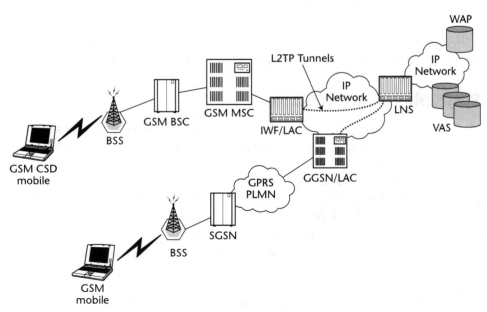

**Figure 6.1** IP VPNs for CSD.

# Packet Data Solutions

Both the GSM and the UMTS systems offer packet data capabilities. The GSM system, which was designed and optimized for the support of speech and circuit data services, has been augmented with packet data capabilities via the General Packet Radio Service (GPRS) overlay. Therefore, the GPRS system does not always deliver optimal data transmission or offer high performance and throughput data services. Conversely, UMTS has been designed from the outset to support packet data services through its PS domain, so its performance is expected to be much more efficient and at a higher data rate than GPRS. Chapter 4 provides an in-depth description of both the UMTS PS domain and GPRS packet data services capabilities. This section focuses on packet-data-based Mobile VPN services that can be provided by both the GPRS and UMTS packet data systems.

The differences in VPN service between the GPRS and UMTS systems are for the most part negligible. The few exceptions are:

- New features introduced in 3GPP Release 99 standards and not present in pre-Release 99 GPRS systems, such as multiple levels of QoS per single data sessions.
- DHCP Relay at the GGSN and Mobile IP FA support at the GGSN.

These features significantly extend the range of services operators can offer, and they will be common to both GPRS and UMTS systems from Release 99 onward. However, it is also expected that after Release 99 packet data services, networks will not be centered around the GSM/GPRS system, and consequently these capabilities are expected to be more widespread in UMTS networks (the exception being EDGE and GERAN deployments, which are unlikely to be a very common in Europe and Asia anyway, and might be limited to other regions, such as North America). For these reasons, this chapter refers to GPRS and UMTS PS domain VPNs interchangeably. When a particular VPN service only applies to some release of the system, we will make that explicit.

## Packet Data Technology Solutions

Let's begin this section by focusing on the GGSN. The GGSN sits between the wireless network and the wireline data networks that interface with it. This network element is common to both GPRS and UMTS systems. This is also the network element that is key to providing advanced data services such as MVPN. The Home Location Register (HLR), the AAA subsystem, Serving GPRS Support Nodes (SGSNs), the user profile, and customer relationship management subsystems are also critical components in the IP services delivery, but the IP services intelligence is concentrated in the GGSN, since this is the point which processes user packets at the network and possibly higher layers. The GGSN is a network element that terminates GTP tunnels established from the SGSN, where the user is located as it moves around the wireless access network. The GGSN provides access points to packet data networks (PDNs). Each access point is identified via a logical name, or the *access point name* (APN). The format of the APN is specified in [3GPP TS23.003], and its syntax has been introduced earlier, in Chapter 2, in relation to DNS usage in cellular systems. The SGSN, at session setup time, resolves the APN via DNS to an IP address or a list of IP addresses belonging to one or more GGSNs offering the desired access point. In fact, for service availability purposes or simply for scalability and load sharing, it is desirable to allow an access point to be distributed over more than one GGSN. This results eventually in the selection of an IP address to be used to establish a GTP tunnel. The actual algorithm used to select the IP address from a list of IP addresses resolved by DNS is vendor-specific. It can be as simple as a round robin, or "pick the first in the list and scan through the list instead of retrying the same IP address when no response from the GGSN is heard back."

The GPRS tunnel establishment process is key to the provisioning of VPN services, and it is explained here in detail. The first message used to set up the GTP tunnel contains the user identity provided by the IMSI and MSISDN. (For a definition of the IMSI and MSISDN see [3GPP TS23.003].) In addition, it carries two other very important pieces of information: the Network Identifier portion of the APN and the Selection Mode. It is possible to authenticate the user based on IMSI or MSISDN—for instance, by passing the IMSI or MSISDN and APN to AAA servers that consider the identity information to be trusted, since it comes from the wireless network. As a part of this process, the GGSN can receive user profile information in the messages received back from the AAA subsystem. Then, this information can be used at the GGSN to retrieve IP service-related parameters and policies from external databases, such as LDAP or COPS directories (see Chapter 5), so that the appropriate policies for the user can be put in place.

The Network Identifier (NI) part of the APN is used at the GGSN to associate the session to an appropriate external network and determine what is the user authentication method and the protocol to be used (IPv4, IPv6, or PPP), as well as whether the PPP session needs to be handled at the GGSN or simply relayed to a LNS via L2TP tunnel (in the latter case, the GGSN would act as an LAC as well). Also, packet handling behavior, policies, external servers IP addresses, host configuration information, and other information can be associated to the APN. A sound GGSN implementation must therefore allow for the configuration of a significant amount of information per APN, to determine in which way the incoming sessions for each APN need to be handled. There are more approaches to service selection and configuration, such as deriving the configuration information from the domain a user specifies together with the username, at logon time. These are discussed later in the chapter.

The Selection Mode information element carried in the Create PDP context request determines in which way the user session is incoming for a specific access point—that is, according to which criterion the user was allowed to use the APN by the network (SGSN). The APN can in fact be MS-specified, network-specified (as in the case of a default APN the user is assigned to by the SGSN), or be a part of the subscription profile and generated by the MS or by the network. Later in this chapter we provide insight on the use of the Selection Mode and the preconditions for it to be useful in providing secure VPN services.

Based on the reception of the APN information, and on lookup of the session-handling information configured for the APN, different network access services can be offered at the GGSN. These network access services can be broadly classified into:

- IP PDP type:
    - Simple IP
    - IP with Protocol Configuration Options
    - DHCP Relay and Mobile IP (available beginning with R99)
- PPP PDP type (available beginning with Release 98):
    - PPP Relay
    - PPP terminated at the GGSN

**NOTE** The current standards are a bit fuzzy in the way they classify the different sorts of network access services offered by GPRS. They oversimplify the classification in "Nontransparent" and "Transparent" access methods (3G [TS29.061] and GSM [TS09.61]), perhaps echoing the original GSM *transparent* and *nontransparent* circuit-switched data bearers. We have therefore chosen to adopt our own classification, which we have just introduced and which will be further detailed in the remainder of the chapter. We believe this to be clearer and more precise than the one currently used by 3GPP standards.

In the standards, *transparent access* is defined to be when the GGSN does not participate in user authentication. The GGSN does not query an external server for user authentication, and the user authentication for network access simply relies on the wireless network access authentication. Wireless network access authentication is performed when the user attaches to Packet Mobility Management (PMM) at the SGSN, or if the user changes SGSN as the user moves, based on trust of information from the old SGSN or based on MS reauthentication at the new SGSN. Broadly speaking, this access method would map to *Simple IP* access mode in our taxonomy (see below the section providing a detailed description of this access mode), should we elect to engage in terminology mapping. But we will show that in Simple IP access mode external servers can be used, and the GGSN can still participate in user authentication.

In this case only the SIM (or USIM) card in the wireless device (MS) is authenticated, rather than the user of the SIM. Even though entering the PIN number at MS startup provides user-level identification, normally users want to skip this phase and configure the MS to automatically remember and use the PIN, thus defeating the purpose of the user-level authentication built into the system. In addition, the PIN is a secret shared with the wireless provider and is not considered a user secret for external network access (that is, the external network cannot base user authentication on the PIN used for wireless access authentication). So, external networks offering "transparent access service" do so based on a trust relationship with the wireless carrier.

In the standards, *nontransparent access* refers to all other access methods where the GGSN participates in user authentication. However, after trying to understand the standards in detail, we are left with many questions and doubts, which often spark lengthy debates on what is transparent and what is not. In fact, PPP Relay over L2TP tunnels appears to be classified as *nontransparent* access, but typically user authentication is performed at an LNS not collocated with the GGSN, so by the definition provided, this should be considered a *transparent* access mode. For the sake of avoiding this kind of interesting but unproductive debate, at least for those who kindly decided to read this book, we opted to use our own terminology, which we consider more appropriate. We also avoided trying to map this to the standards taxonomy, since again this would have required us to justify decisions taken in the standards, which appears to be quite a risky and difficult task and not necessary for the readers to fully master the matter and properly apply these solutions in the real world.

## IP PDP Type

The IP PDP type allows for the provision of IP network access services for both IPv4 and IPv6 by offering IP layer connectivity and services to the MS. We have decided to consider IPv4 only in this chapter for the sake of simplicity, since it will be mainstream for a few years to come in the provisioning of corporate network access services and advanced IP services.

Solutions based on this PDP type encompass different ways to offer IP address assignment, host configuration, and lower-layer connectivity to the IP network. The value of the network identifier portion of the APN sent to the GGSN in the *Create PDP context* request determines which combination of these service building blocks will be used for the sessions based on GGSN configuration. Also, other information can be provisioned at the GGSN on a per-APN NI basis, such as the next hop for uplink packets (in case this helped to route packets to appropriate destinations on a per-APN basis, such as in the case where a different ISP or network is associated with different APNs).

### Simple IP

An APN configured for Simple IP mode of access offers the following types of services:

- Layer 2 (ATM, MPLS, Frame Relay, PPP, etc.) or tunnel-based (IPSec tunnel mode, IP/IP, GRE, etc.) connectivity to the external network.
- Possibility to interface to a AAA server to perform an IMSI- or MSISDN-based authentication or RADIUS-based IP address assignment.
- Use of RADIUS accounting to communicate session-related events to accounting servers or application servers.
- Dynamic or static IP address assignment.
- Network-initiated PDP context activation.

When network-initiated PDP context is supported, the IP address needs to be statically associated to the IMSI of the MS. The IP address is assigned using local pools at the GGSN or RADIUS or DHCP client, and it is communicated to the MS in the end-user address IE of the GTP *Create PDP context* response and RIL3 [3GPP TS24.008] *Activate PDP context* accept messages.

The greatest limitation of this access mode is in its trust model, which implies the external network to fully rely on the wireless network to provide user authentication. There is neither a human-kept secret (password) nor two-factor authentication (human-kept secret plus a token-card-generated one-time code) that can be used to prevent individuals from getting hold of a terminal, by accident or maliciously, to access the network associated to the APN, if the true owner of the terminal disabled the need to insert a PIN to get the MS attached to PS services. For this reason, this access mode is most useful for providing access to applications and services that do not require user authentication. In addition, if an application accessed via Simple IP requires strict user authentication for sensitive transactions to be carried out, some login- and password-based authentication within the TLS session can be used.

On the other hand, this access mode is most suitable for services that require minimum interaction between the user and the terminal in order to set up connectivity. This access mode, if accompanied by the usage of RADIUS Authentication or Accounting, can also be used to provide single sign-on by transferring session-related information to a services access layer that distributes the user identity and IP address used for mapping to applications. In fact, the service access layer may "know" of the mapping of the IP address to IMSI or MSISDN via RADIUS-based IP address assignment or via reporting of the mapping of the IP address to the user identity (IMSI or MSISDN) via RADIUS accounting messages. Today this IP address-to-user ID mapping is mostly used in WAP gateways or HTTP proxy servers to enable advanced content conditioning and billing features.

Normally, Simple IP can be expected to be used for Web- or WAP-based browsing applications. Another possible application of this network access mode is end-to-end VPNs—that is, client-based remote network access. Recall from Chapter 5 that the VPN client-based network access requires public routable IP addresses to be assigned—unless nonstandard-compliant IKE implementations are adopted. The latter condition, however, requires the service provider and the corporations to agree on which VPN client and gateway vendor to pick, which does not appear to be practical in most cases. Recently, some IPSec NAT traversals proposals to the IETF make the use of private addresses possible for IPSec tunnel mode operation, thus relaxing the constraint to have public IP addresses assigned.

In Simple IP the GGSN can be provided with evidence the subscriber to the APN has access rights based on the *Selection Mode* Information Element carried in the Create PDP context request, if the selection mode attribute is set to value 0—that is, "MS or network provided APN, subscribed verified." Of course, if reliance on this information is fundamental for the proper operation of the service, the GTP signaling must be protected by security measures that would enable integrity preservation. Alternatively, the GGSN can query an AAA server with *RADIUS Access Accept* populated with the user MSISDN or the IMSI RADIUS 3GPP VSA (as described in [3GPP TS29.061] and in Appendix C), so that it would be possible to perform authentication of the user based on the IMSI or MSISDN trusted information provided by the network. In this case, the *Access Request* username and password would be populated with some dummy values.

Again, the IMSI or MSISDN information transported by GTP signaling should be protected, so that its integrity (and, if desired, its confidentiality) can be preserved. Typically this is achieved by using GTP protected with IPSec.

The IP addresses can be assigned in this case via local address pool, RADIUS, or DHCP client. Static IP addresses are also allowed, and in fact, the use of static IP addresses is necessary for network-initiated PDP context activation.

The host configuration is yet another area in which this solution is not as strong as the others. With Simple IP the MS is required to be manually configured by the user, possibly with the aid of some software tools provided with the subscription pack on an installation CD-ROM, with the IP address of the NetBIOS (Network Basic Input-Output System) servers or DNS servers to avoid headaches to the service provider's help desks. This may not be desirable if no automated and constrained mechanism is in place. In summary, this access mode is suitable for simple terminals requiring access to applications that can resolve strict *user* authentication in a way independent from *network access* authentication.

### IP with Protocol Configuration Options

IP with Protocol Configuration Options access method architecture is depicted in Figure 6.2. The *Create PDP context* request message can contain the Protocol Configuration Options Information Element (PCO IE). This Information Element is a transparent container of host configuration and authentication information that is exchanged between the terminal equipment (TE) and the Mobile Terminal (MT) components of the MS. The TE component is expected to be a laptop or another device that interfaces to the MT via a PPP-based link. The PPP authentication phase can be based on "No auth" PAP or CHAP. The MT *always* successfully authenticates the TE, collects authentication material from it, and enters the Internet Protocol Control Protocol (IPCP) phase. This authentication material and IPCP configure request is then included in the PCO IE in an *Activate PDP context* request sent to the SGSN, which then transparently transfers this information to the GGSN in a *Create PDP context* request message. Eventually, the GGSN uses this information to authenticate the MS. If the MS is authenticated, then the GGSN also determines which host configuration information needs to be sent back to the MS (including an IP address for the MS, the IP address of the primary and secondary DNS servers, or possibly the IP addresses of the primary and secondary NetBIOS name servers) using a PCO IE in the Create PDP context response.

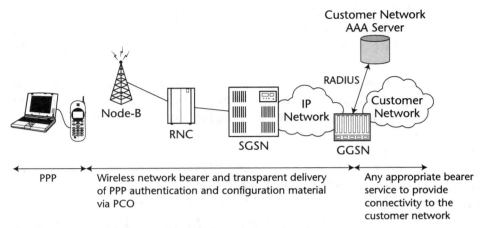

**Figure 6.2**   IP architecture with PCO-based access mode.

This access mode based on IP PDP type allows for the same layer two or tunneled connectivity to the network associated to the APN as in the Simple IP case, but it adds the capability to perform user authentication for network access based on a secret shared between the administrative entity of the external network and the end user, thus allowing for a stricter level of security than Simple IP mode. The only weakness in this model is the possibility for a malicious user to snoop a challenge/response pair sent in a PCO IE and then reuse that to gain access to the network. In fact, this network access mode does not allow the GGSN (or the AAA subsystem) to generate a challenge to the MS, thus exposing the system to replay-based attacks. In general, a well-designed network should not be exposed to such attacks, unless the malicious user can craft a handset to generate this stolen challenge/response pair, which is generally not a trivial task. Network administrators still concerned about the weakness associated with replay-based attacks can always turn to the PPP PDP type, in relay or PPP terminated forms, for assistance (see the section "PPP PDP Type" coming up in the chapter for details).

By defining the username to be compliant to [RFC2486] format, in which the concept of the NAI is introduced to define a username with the format "user@domain," IP with PCO access mode allows for the GGSN *in a visited network* to be used, provided that AAA level roaming capability is in place (à la iPass and GRIC—two ISPs offering global Internet access based on roaming agreements with other ISPs in different countries). Also, by changing the domain component, some intelligent IP services platforms can be configured to return in the filter ID attribute or other RADIUS attributes the name of a service whose definition, in terms of network access policies, can be retrieved from an LDAP or equivalent service policies configuration data repository. The different service policies may, for instance, allow the GGSN to route user packets to different networks depending on the domain component of the username, thus allowing a subscriber to select a specific network and the service it offers based on this value.

Also, by adding the "3GPP-GGSN-MCC-MNC" RADIUS Vendor-Specific Attribute (VSA, see [3GPP TS29.061]) to the RADIUS messages, when a GGSN in the visited network is used at the home AAA server, it is possible to apply visited-network-dependent policies. The AAA subsystem may as well trigger applications in the home network to send to the MS visited network-specific push content, such as local news or alerts. The AAA server in the home network may in fact instruct push servers in the home network to initiate push sessions with the MS, using the address that is included in *Accounting Request START* that has been received from the GGSN. Alternatively, if the GGSN is in the home network, an equivalent functionality is provided by the usage of the "3GPP-SGSN-IP address" RADIUS 3GPP VSA to determine whether the user is "at home" or roaming. With Reverse DNS lookup it is also possible to retrieve additional information and identify the current provider or determine geographic location information.

### DHCP Relay and Mobile IPv4

Release 99 of the 3GPP standards has enhanced GPRS specifications to allow an APN to be configured to support DHCP Relay service or Mobile IP Foreign Agent (FA) functionality. Figure 6.3 depicts a typical DHCP Relay access method scenario. When the *Create PDP context* request is sent to the GGSN for an APN configured to support DHCP or Mobile IP FA, a *Create PDP context* response is sent back to the SGSN immediately without any other user authentication than in Simple IP access mode. This defines a GTP tunnel and a bearer toward an MS without any MS IP address associated with it. This tunnel can be used to exchange DHCP configuration messages or Mobile IP advertisements and registration messages. Eventually the MS will be assigned an IP address using DCHP or Mobile IP methods. The remote network access will be obtained using encapsulations methods allowed by Mobile IP, or by using the link layer and tunneling technologies defined for Simple IP when DCHP is configured.

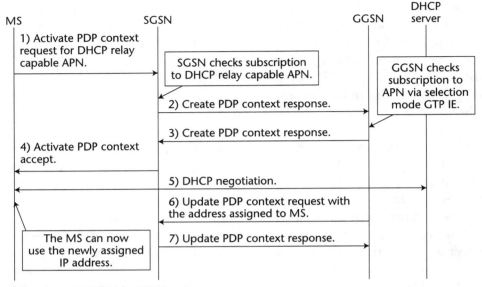

**Figure 6.3**   DHCPv4 in GPRS systems.

The DHCP Relay access mode is likely to be used when advanced host configuration methods and when a "LAN-like" access mode are required. The LAN-like access mode is particularly suitable for wireless devices that require discovery of a lot of service-related information, such as the HTTP or SIP proxy IP address. Generally, the user authentication in this method suffers from the same shortcomings as in the Simple IP mode. However, here user authentication may be further enhanced by the use of the DHCP authentication [RFC3118].

The Mobile IPv4 access mode is also suitable for LAN-like access mode, as it seamlessly supports the handoff between GPRS/UMTS and other access technologies such as Wireless LAN. Note that Wireless LAN-based access is key in many hot-spot-based deployment scenarios (more in Chapter 9) and is appealing for its high data rate and the low cost per byte. The Mobile IP-based, combined GPRS/UMTS/WLAN networks may become widely deployed in the future, once some security and standard issues have been resolved and when interoperable and affordable user devices are available.

## PPP PDP Type

The PPP PDP type was added to the GPRS specification beginning with Release 98. This was a very important addition to the capabilities offered by the GPRS system in that it allows more seamless adaptation to the installed base of wireline remote network access infrastructure that is today in most cases based on PPP. It also solves weaknesses in the CHAP implementation based on the IP with Protocol Configuration Options access mode, as detailed previously. PPP PDP type enables the use of PPP encryption and PPP compression, as well as the use of network layer protocols other than IP. PPP also defines an Extensible Authentication Protocol (EAP—[RFC2284]) that allows the LCP negotiation to terminate without determining the authentication protocol, which is transparent to the NAS and only determined at authentication phase time. This allows for the evolution of the authentication protocols used without the need of changing the NAS and AAA infrastructure. It also allows for the use of advanced authentication algorithms that will be developed over time, such as smart cards and biometrics, that cannot reuse existing authentication methods such as PAP and CHAP as authentication information transport method.

PPP periodically checks the availability of the end-to-end link by using an LCP *echo request/response* message. This can lead to a number of problems related to bringing up the radio links even when no useful data needs to be transmitted. Here is how this problem can be eliminated: Both the GGSN and the MT have local GPRS/UMTS bearers availability information. So both entities can avoid relaying LCP echo requests and just respond to the echo requests themselves. In the PPP Relay case, both the GGSN and MT will act as LCP echo message proxies (the GGSN toward external NASs, the MT toward the TE). When PPP is terminated at the GGSN, the GGSN should not transmit LCP echo requests, and the MT should act as an LCP proxy. This setup guarantees optimal performance of a PPP PDP type based MVPN, and it does not represent any practical limitation in detection of the link state.

Some MVPN client implementations—such as L2TP-based VPN clients and IPSec VPN clients—normally exchange *keep-alive* messages with the VPN gateway. In this case the network has no control over them, nor can it act as a proxy to avoid inefficient usage of radio resources. Therefore, it may happen that this would negatively impact the radio resources usage and make the bill for the user undesirably more expensive. Also, a PPP PDP type based solution with LCP proxy would allow the end-to-end bearer to be up as long as the wireless bearer is up, whereas a VPN client-VPN gateway link may be down even when the wireless bearer is not (for example, because the VPN tunnel keep-alive messages were lost over the radio). Because of these deficiencies and others described in Chapters 4 and 5, it appears clear that the end-to-end, VPN client-based solutions often may not be optimal in cellular environment, both from an operator and from a wireless subscriber perspective. Therefore, compulsory MVPN solutions built around properly executed PPP PDP type or IP PDP type access methods are more likely to be successful in the cellular environment.

The additional benefit of a PPP PDP type based MVPN is that in the PPP Relay case, the service provider can let the private network administrator perform address management and AAA, thus minimizing the impact on the cellular network administration and complexity. On the other hand, operators can offer outsourcing of these services and also consolidate on a unique platform terminating L2TP tunnels coming from both CSD and PS bearers-based access—and even dial, broadband, and wireless LAN access.

### PPP Relay

In PPP PDP access type an APN can be configured to relay PPP frames to a predefined external NAS device. The de facto standard technology used in this case is L2TP (see Chapter 2 for background on L2TP). L2TP can be relayed over Frame Relay, ATM, and UDP/IP. The APN at the GGSN must be configured with the L2 address (Frame Relay or ATM) or the IP address of the LNS, as well as the L2TP tunnel name and password. The information associated to APN determines which remote network PPP frames are going to be relayed to, thus requiring the GGSN only to set up the tunnel and the L2TP calls within it. This constitutes a really simple setting that can also guarantee a sufficient level of end-to-end security when the L2TP tunnels are secured via IPSec transport mode and PPP encryption is negotiated. Also, as it can be anticipated, in this scenario the GGSN may be acting as an LCP echo proxy. Figure 6.4 depicts the protocol stacks involved in a typical PPP Relay configuration using L2TP transported over UDP/IP.

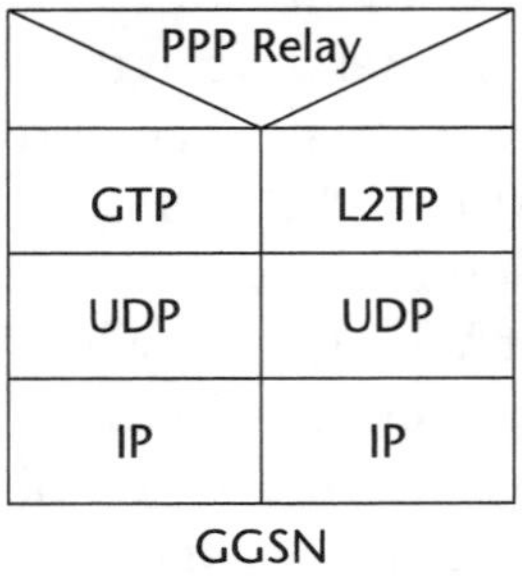

| PPP Relay | |
|-----------|-----------|
| GTP | L2TP |
| UDP | UDP |
| IP | IP |

GGSN

**Figure 6.4**  PPP relayed using L2TP.

Since the GGSN transparently attempts to set up calls to the LNS configured for the PPP Relay APN, it is recommended to include the APN in the set of the PDP context information stored in the HLR. This way, the incoming selection mode IE in the Create PDP context request is set to value "0"—or "MS or network provided APN, subscribed verified"—and those subscribers who are not authorized to attempt to set up the L2TP tunnels are denied the right to perform L2TP call setup. This feature helps protect against denial-of-service (DoS) attacks. Also, the Calling Number L2TP AVP should be set to the MSISDN of the MS, so that the LNS may be configured to reject incoming calls from calling numbers that do not belong to a certain preconfigured set of allowed numbers. The LNS administrator may use this option to detect the MSISDN of users attempting to access the LNS without rights, if necessary, for security reasons. Also, sending this AVP is necessary for the LNS to relay the MSISDN information to the AAA subsystem or to WAP gateways via a proprietary or RADIUS-based interface (in which case the RADIUS attribute used would be the Calling Station ID attribute).

### PPP Terminated at the GGSN

PPP terminated at the GGSN access method offers the benefits of the PPP protocol-based host configuration and authentication in addition to a very flexible network access paradigm allowing for an array of advanced IP services. For instance, when the user is authenticated, the AAA server can return a name of a service to be provided to the user (as described in the previous section, "IP with Protocol Configuration Options") or possibly the information necessary to tunnel PPP frames to a LNS. The GGSN can also support PPP compression (typically LZS and MPPC), which allows for more efficient use of radio interface.

On the same GGSN platform used to terminate PPP PDP type GTP tunnels, it is often possible to terminate/originate L2TP tunnels, and therefore consolidation of multiple access technologies, both wireline and wireless, is also possible. Figure 6.5 illustrates in particular the protocol stacks supported by a GGSN terminating PPP.

A comparison between PPP terminated at the GGSN and IP PCO access modes will help us assess the weaknesses and strengths of each of them. The PPP terminated at the GGSN mode is more friendly to the GTP protocol operation, since in this case the GTP tunnel can be established immediately, without requiring the GGSN to wait for the user AAA and configuration processes (and possibly L2TP tunnel establishment, when the tunnel attributes are passed back in RADIUS Access Accept) to complete.

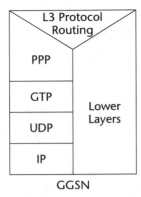

GGSN

**Figure 6.5** PPP terminated at the GGSN.

In some GGSN implementations, it is also possible to set up the GGSN to immediately establish an L2TP call when an incoming IP PDP type Create PDP context message is for a specific "IP with Protocol Configuration Options" access mode APN. This setup, however, would constitute a non-standard usage of L2TP, and it makes the end-to-end session vulnerable to the potential replay based attacks that affect the IP PCO mode. The L2TP tunnel establishment and user AAA process may also take long enough to create problems for GTP protocol handlers at the SGSN. In principle, an operator may tune the GTP timers and retransmission attempts for create PDP context requests to allow for the latency associated to "IP with Protocol Configuration Options" in setting up tunnels, but this is not a generally safe measure and also does not offer sufficient guarantees to provide acceptable service also when the user is roaming to networks that do not adopt the same tuning of GTP parameters.

Having said this, the solution might address the current shortage of PPP-capable GPRS terminal implementations that is affecting the industry, while still allowing the service flexibility offered by the other approach. Finally, the PPP PDP type allows for PPP compression protocols, such as STAC LZS and MPPC, to be used and negotiated, which is not something IP allows for. This can make the extra overhead added by PPP (2 bytes per packet) totally insignificant. As a summary, IP with Protocol Configuration Options results to be dominated by the PPP PDP type terminated at the GGSN, but it still can be expected to survive for quite a while at least until PPP PDP type support in terminals is widespread.

# Service Level Agreements

Service level agreements (SLAs), which a UMTS Mobile VPN service provider defines with a customer, include business arrangements and financial and legal clauses that are not relevant to technology and hence beyond the scope of this book. Usually, SLAs include availability figures, packet loss per class of service, replacement policies of failed units in the customer network if the operator also provides the customer premise equipment, troubleshooting and help desk support to administrators, technical training for administrators, IP addressing information, and the scope of the variables the customer can remotely manage.

The availability and support commitments agreed in the SLA can be expressed in terms of mean time between failure (MTBF), mean time to repair (MTTR), and reachability of technical support or availability of spare parts to replace failed parts. For instance, there can be different tariffs applied whether continuous support or limited support is guaranteed.

The guaranteed QoS levels could also be part of the SLA, together with a traffic conditioning agreement according to the DiffServ model (see Chapter 2), including a traffic profile a customer must comply with and the policing and remarking rules a service provider would enforce at the boundary with the customer network to traffic complying and not complying with the traffic profile. The SLA also specifies how IPSec sets up security and confidentiality features, including:

- Which encryption and message header authentication algorithms are expected to be used

- Whether manual configuration or PKI infrastructure is used for authentication keys distribution

- Whether tunnel or transport mode is used

- Particular IPSec policies

- The IP addresses of the security gateways

Password management criteria for L2TP tunnels should also be included. Within this security-parameters-related section of the SLA, the handling of subscriber profiles and data should be described. Also, the trust relationship existing between the customer and the provider often depends on very specific clauses and guarantees written in this section.

The account setup and service sign-up methods for subscribers associated with the customer network must be part of the agreement. The service provider may provide a service signup Web page for this purpose. The type of subscriber authentication information that can be requested in order to obtain troubleshooting and support or benefit from customer care services must be included, and the handling of such data, that is confidentiality and privacy matters, must be spelled out.

Other specifications the SLA may include are:

- AAA server access method (via proxy or direct access or broker network), as well as IP addressing information for host configuration information servers and network access methods allowed (IP with PCO, PPP Relay, PPP terminated), together with the availability, security, and AAA message attributes required for the service delivery.

- Billing date and payment methods conditions, integral usage data documentation, and other billing and financial aspects.

- For the CSD case, NAS telephone numbers, as well any conditions associated with user sessions termination upon idle timeouts.

- IP addresses for LNS or other tunneling protocols endpoints.

- Availability of the MVPN service when roaming and associated roaming fees.

It is not our intention here to provide an exhaustive list of what an SLA for MVPN may include (especially considering that some of these details had been covered in Chapter 5); however, we do once again want to stress its importance. This document, beyond the legal and business aspects, sets the *customer expectations* and also defines the service the customer will ultimately receive. So it is important that both the service provider and the customer can perceive it as a useful tool for their interaction, service definition, and implementation.

Also, since there will likely be a large set of customer network requirements, the level of customization of the SLA might vary widely depending on the size of the customer MVPN. It also depends on whether the provider wants to standardize a service or whether the provider wants to use the flexibility of its network to accommodate different customer needs.

## Charging and Billing

If the expectations of MVPN services to become one of the mainstream cash generators for wireless service providers are ever realized, accounting data collection and billing information generation will surely become some of the most critical aspects in the successful and profitable delivery of

MVPN services. Operators may define charging plans based on tariff times, traffic volume thresholds, location, or other parameters, such as application-level information derived by deep packet inspection. All charging plans are ultimately based on an appropriate charging data collection. Prepaid access-based subscription plans are also possible. Although this is certainly not the most common scenario for data VPN for corporate network access, it will play a significant role if access to application-specific VPNs is used to support consumer services.

The GPRS charging architecture was outlined in Chapter 4. GPRS charging is based on the Charging Data Records (CDRs collected for wireless access usage accounting. However, RADIUS accounting is also used to account for session duration and possibly to interface with an accounting infrastructure operated by a partner network. For example, RADIUS is used when the customer network requires collection of accounting data for trend analysis and usage profiling, and possibly of charging for network access itself in a manner independent of the billing performed by the wireless service provider. The standards also define the support of prepaid services in GPRS. The standard is CAMEL Phase 3 ([3GPP TS23.078], see Figure 6.6), which defines the interaction of the SGSN with the GSM SCF for the provision of prepaid services. The protocol used for such interaction is called CAMEL Application Part, or CAP, defined in [3GPP TS29.078].

## Roaming

One of the strengths of the GSM/GPRS and UMTS systems is its seamless roaming capability across countries and networks belonging to different operators. This capability can also be used to support Mobile VPN services.

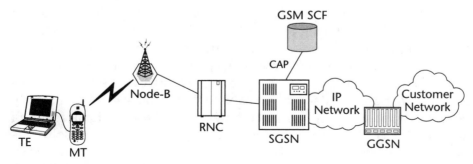

**Figure 6.6** CAMEL Phase 3 prepaid systems architecture.

Roaming support in GSM has been at the core of the reasons why the GSM Association was established in the first place. Many operators (mostly PTTs, for Post, Telephone, and Telegraph, in the early days) agreed to provide service to subscribers roaming into their own networks from other Home Public Mobile Networks (HPMNs) according to a well-defined set of rules set forth in the GSM Memorandum of Understanding (MoU). This has spurred a number of activities within the GSM Association International Roaming Expert Group (IREG) that helped in fine-tuning technical details in providing roaming to subscribers across networks, across countries, and for different services.

One of the GMS MoU's guiding principles is that a Visited Public Mobile Network (VPMN) cannot provide more services than the user has sub-scribed to in the HPMN. Networks entering into a roaming agreement need to specify what services the roamers are entitled to receive when in visitor mode, and also have to agree to rules governing the ways such services might be denied. The home operator may always define classes of users that can be offered roaming service by VPMNs by defining barring information on all or a subset of the services available in a VPMN. This information is stored in the HLR and downloaded in the visited network serving node at the time of user attachment, or it is transferred to a serving node when a user performs a location update procedure/handoff. When the mobile station or user equipment attempts to attach in a roamed-to net-work and the user has no right to get roaming service, the network may signal this, and the mobile station will not attempt again to attach to the network.

Roaming was initially conceived to enable service across countries, but as soon as market deregulation took place, regulators forced incumbent oper-ators to provide access to their own infrastructure to new entrants on the basis of regulated roaming rates to alleviate entry barrier problems and pro-mote competition on a fair basis. This scenario is also being repeated by 3G, except in this case, greenfield operators (that is, new companies starting up 3G businesses without having been a player in 2.5G GPRS network deploy-ment) waiting to roll out 3G service are currently trying to establish some customer base by defining GPRS roaming agreements with incumbents.

Because of the high importance of roaming, the rest of the section focuses on enabling roaming for data services. Note that so far we are at the early stages of the wide-scale deployment of international and national roaming capabilities we could expect for the future. Existing roaming solutions are limited in scope, and no significant commercial deployments have been recorded at present nor are expected in the short term. Also, the standards for CAMEL still have some ambiguities that make interdomain, multivendor operation of prepaid mechanism not likely to happen very soon, mostly because of interoperability problems (which are being sorted out at this writing).

For the CSD case, roaming support is all but equivalent to the roaming support for voice services. In this book we won't describe the details of CS voice capability, except for the need for users to have a list of phone numbers available in visited networks that would allow them not to dial back to access numbers in the home network. This may be accomplished according to the IETF ROAMOPS (for Roaming Operations) WG guidelines or manually in a proprietary way. It should be noted that WAP services access in roaming conditions could be made far more efficient and cost-effective if carriers provided POPs around the world by defining a phone list users could use as they dial from around the world. Unfortunately, though, carriers are concerned about phone book configurations on terminals and also have shown reluctance to deploy advanced roaming solutions before the service has really taken off within the home network. This reluctance has, in essence, stopped the deployment of such a solution to date.

GPRS/UMTS data roaming is governed by both standards and industry consortia documents such as [PRD IR34] from the GSM Association IREG. The GPRS standards allow for a user to roam in a visited network and use a GGSN in the home network, or use a GGSN in the visited network. The interface between the GGSN in the home network and the SGSN in the visited network is called the Gp interface (see Chapter 4). The GTP tunnel, when a GGSN in the home network is used, traverses a network that is provided by a transit network service provider. This network, depicted in Figure 6.7, is called GRX (GPRS Roaming Exchange), and according to IREG guidelines, it is based on a public addressing scheme.

Access to the GRX may happen at central exchange points, which are equivalent to IXC or Internet Exchange Points, where many operators can exchange roaming traffic and set up BGP4 peering, over an L2 infrastructure provided by the GRX provider. The BGP routes advertised over the GRX are not distributed outside the GRX itself, and also no Internet routes are distributed over the GRX. Therefore, there is no mutual network layer reachability between the Internet and the GRX: the GRX is a private network based on a public addressing scheme. This was believed to be the best option because IREG members assumed that coordinating the use of a private address space between operators would be unpractical. However, the operation of a GPRS network may require a significant number of public IP addresses and the Internet Addresses registries are known not to give out many of them lately, so this may prove to be a hurdle in network operation.

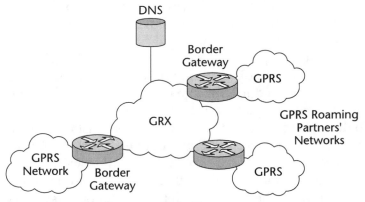

**Figure 6.7**  GPRS roaming networking.

**NOTE** Typically, a GRX also offers DNS root service for the .gprs domain so that resolution of access point names to IP addresses in remote networks is possible from any GPRS network.

GPRS-based MVPN service is normally offered based on a GGSN in the *home network*. This is accomplished normally by devoting an APN to the customer network, and this APN will resolve to an IP address or a list of IP addresses belonging to a GGSN in the home network. This approach requires the GTP tunnels between the SGSNs and home GGSNs to be protected via IPSec transport mode, so that the trust relationship between the visited network operator and the home network operator is not required to be extended over all the network providers traversed by the GTP tunnel. However, it is also possible to use a GGSN in the visited network, by defining a well-known APN that can be translated by the SGSN of the visited network into an APN that resolves to one or more IP addresses belonging to GGSN in the visited network. This requires a PPP PDP type APN, or an APN supporting the IP PDP type with PCO access mode, and the GGSN ability to dynamically assign the incoming request from the user to an appropriate VPN and to establish connectivity if no connectivity is statically set up. User assignment to a VPN is usually based on user profile information retrieved from the AAA subsystem (e.g., via the Filter ID

RADIUS or "RADIUS L2TP Tunnel Information" attributes. Other solutions may require more customization of the GGSN node (e.g., lookup tables).

When a user is roaming using a home GGSN, accounting information at the GGSN is critical to record the usage data in the home network independently of the visited network. Also, the home GGSN, as mentioned, may use RADIUS accounting to accommodate the customer network's needs. User authentication in a home GGSN scenario is performed exactly as in the nonroaming scenario. For cases that rely strictly on the HLR subscription data for user authentication, integrity protection of GTP signaling from the visited SGSN to the home GGSN is required, so that the Selection Mode IE cannot be altered and therefore is expected to be valid, since it is generated by a visited network that has a trust relationship with the home network. As a part of the roaming agreement, the way the GTP signaling integrity is guaranteed may be subject to negotiation and definition. By the same token, IPSec polices for VPNs may be defined as a part of the roaming agreement.

User authentication in a visited network GGSN is normally governed by a AAA roaming agreement, where the visited GGSN may be acting as a AAA client to an AAA infrastructure based on RADIUS proxy and possibly including a RADIUS broker (see Figure 6.8). However, this kind of arrangement is likely not to dominate in the GPRS world, unlike the CDMA2000 networks that strictly rely on this setup (see Chapter 7).

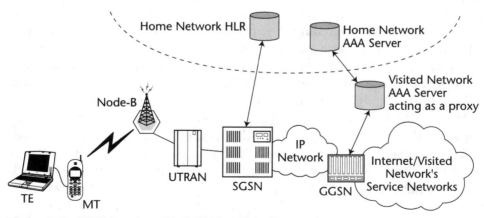

**Figure 6.8**   GPRS roaming with GGSN in a visited network.

# Case Study: ACME Wireless

Operators in the GSM and UMTS arena can be segmented into four broad categories:

**Incumbent.** These normally originated from PTTs or a regulatory-driven breakup of a monopoly, or by one of the companies obtaining a second license in European countries.

**Latecomer incumbent.** Normally not "in the black," as far as their current balance sheet is concerned, these are still players in the 2G market.

**Greenfield.** These are operators who enter the market as pure 3G players. They are already financially up and running but are yet to have an operating network.

**Mobile Virtual Network Operators (MVNOs).** These are operators that do not own a wireless access network but simply rely on roaming agreements with incumbent cellular wireless providers in order to provide service to the customers who choose to use their brand. In brief, a MVNO focuses on service definition and branding and on the highest-quality-possible customer care, and it negotiates with a third parties wireless access network services (more on MVNOs in Chapter 9).

In this case study we focus on a provider that belongs to the first class, named ACME Wireless. This carrier type provides the most complete and simple arrangement in terms of networking and service provisioning, since the latecomers often have to use the incumbent's infrastructure and set up complex service arrangements based on roaming agreements even within their home country. Greenfield and MVNO cases may be considered a subset of what can be developed in an incumbent network, without the need of legacy investments protection but with the need of detailed SLAs and partnerships with other wireless carriers, respectively.

Operator ACME Wireless has offered CSD-based services for some years, namely plain Internet access and WAP services. They have also rolled out GPRS and plan to support more advanced data services as they move toward 3G. ACME Wireless infrastructure is depicted in Figure 6.9. ACME Wireless currently has a TDM network for digital voice transmission and a multiservice network based on ATM and Frame Relay-capable switches. The ATM backbone is used to interconnect their five regional offices and the data center where they host their mobile Internet services (WAP gateways and servers, Web hosting and caching, HTTP proxy) and also to connect the GGSNs located in data centers and the regional offices to POPs where ACME

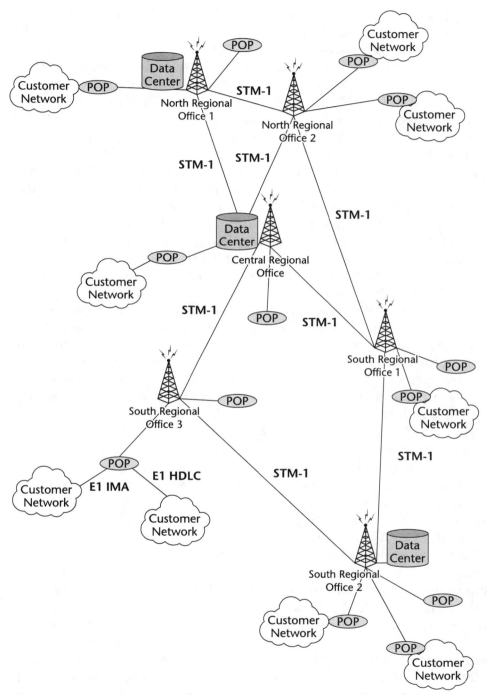

**Figure 6.9**   ACME Wireless network.

Wireless terminates the customer networks' access links. The regional offices and the data center host the provider edge (PE) equipment. The POPs also provide outsourced PE L2-based or PE L3-based connectivity to the customer networks. The POPs are outsourced from major wireline operators operating landline VPN service. This outsourcing service is standardized to be based on MPLS VPNs compliant with [RFC2547]. Each customer can access the operator network either based on a Frame Relay or T1/E1 IMA (Inverse Multiplexing ATM) link. The wireline operator aggregates this traffic at POPs onto the MPLS VPNs they offer to connect multiple sites of the same customer network, as well as to offer connectivity to the wireless network by interfacing to the ACME Wireless regional centers.

The network evolution entails migrating to a unified IP-based voice and data network. The transport will be MPLS based. ACME Wireless plans on reusing the ATM network by interworking the MPLS layer and the ATM layer at ATM edge nodes, to maximize reuse of the existing installed base. Exchange of traffic with the corporation may be based on L2TP tunnels secured using IPSec transport mode or on tunnel mode tunnels. This provides ACME Wireless with the maximum flexibility in selecting partnerships in offering POPs to customers. In fact, secure tunnels decouple the VPN provisioning architecture from the link layer access technology and from any mutual trust between the wireline access operator and ACME Wireless.

If the customer network is outsourcing remote access to some wholesale access provider, then ACME Wireless can cover this need from the wireless side by terminating PPP at the GGSN, or by using the IP with PCO access method. ACME Wireless advises their customers that IP with PCO access mode may be prone to replay-based attacks, and that PPP-based access is best for security and for the other reasons we have suggested earlier in the chapter. In case the customer network uses remote access via L2TP, the ACME Wireless can provide LAC functionality via the PPP PDP type. There are solutions based on IP PDP type and LAC functionality obtained by having the GGSN initiate L2TP tunnels and manage all the PPP negotiations and configuration, using data transferred to the GGSN via GTP. ACME Wireless does not believe this is the target solution, but it still offers this option to customers not equipped with PPP PDP type-capable terminals. In fact, PPP PDP type has not been widespread in the early days of GPRS and UMTS system deployment, but it is becoming more and more widespread as the infrastructures mature. It should be noted that ACME Wireless did not accept the proposal from some vendors to base the whole Mobile VPN offer on VPN clients installed on terminals, since this is a less profitable approach and it does not allow for service delivery control to the same extent as other network-based approaches. For example, the provider can control PPP LCP echo via proxy at the GGSN or by disabling it at the

GGSN when PPP is terminated at the GGSN. Keep-alive messages generated by VPN clients cannot be controlled, since this is perceived as regular user traffic by the infrastructure. Also, IP PDP type network-based VPN solutions do not generate periodic keep-alive messages over the air. This allows for long inactivity periods without requiring the wireless bearers to be allocated permanently to the wireless user. As a result, ACME Wireless plans on network-based solutions only.

The option to manage outsourced IPSec VPN gateways for their customers is also perceived as costly, with no apparent benefit. In addition, it requires the VPN GW/VPN clients to be standardized for the network, because of common interoperability issues between VPN clients and GWs from different manufacturers. ACME Wireless perceived this as suboptimal during its development plans definition.

ACME Wireless also must protect legacy investment in data services and the customer base. Typically, this is represented by CSD-based WAP services users. In fact, plain remote access did not require corporations to set up any agreement with the carrier, since the access number to be dialed was the same as the one used for wireline access. The low bit rate and limited use of service led to almost no subscribers to the L2TP-based access service, offered by having the IWF acting as LAC. The real worry was to make the WAP infrastructure as common as possible between the CS and the PS domains. This was fairly easy because of the uniform way to access WAP services from GPRS and CSD offered by reusing the same WAP GW and WAP GW interaction procedures, achieved via L2TP access to an LNS interacting with the WAP GW.

ACME Wireless also plans to offer to ASPs access to a set of services that enable users to charge for application usage out of their wireless bills. In addition, it plans to extend to users network usage discounts based on the purchase of goods and services or the usage of specific applications offered by ASPs, as a means to control the amount of traffic exchanged between the carrier and the ASPs. The network offers APIs to partners (ASPs) to program their services, as well as dedicated APNs to access partner networks services.

ACME Wireless also can offer single sign-on service to customers by collecting session-related data using RADIUS accounting at the GGSN, or via captive portals (described in Chapter 5) that collect the user identity and then manage it across the applications usage workflow. The user identity can be distributed to partners via some HTTP request URL modifications operated by content switches that allow the identification of the user during the session. The user identity passed in the HTTP requests can then be used by the partner ASP to request network services like billing, discounting, location, or even user preferences and class. The single sign-on service comes also with the recall of all user preferences and bookmarks to be used during the navigation. Figure 6.10 depicts the architecture that enables ACME Wireless to offer such services.

From a network management and service provisioning perspective, the integration of the applications with the network elements configuration is in the plans, obtaining real *flow-through* provisioning (Figure 6.11 shows an example of a possible flow-through provisioning setup, and much more complex scenarios are also possible). This would, for instance, allow a user to begin a session with a simple APN devoted to services network access and then connect to the corporate network, or to a gaming network, where a community of people can share information or exchange media over a specific and predictable QoS-level network. At each stage, the cost of network access would change, thus allowing for dynamic adaptation of the network access fees to the application being used. This could come with advantages to both ACME Wireless customers and to ACME

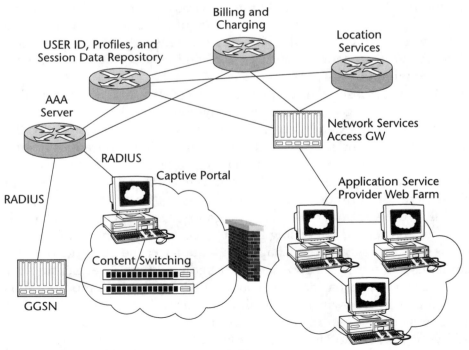

**Figure 6.10**  User identity management architecture.

Wireless themselves: Customers are charged a fair price for each activity they perform, while ACME Wireless retains customers and appeals to new customers with fair rates, offering a predictable application environment and receiving the appropriate revenue for each network access service offered.

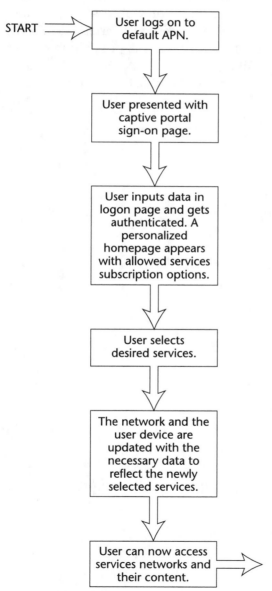

**Figure 6.11**   Flow-through provisioning.

## Summary

In this chapter we discussed how MVPN can be offered in GSM/GPRS and UMTS environment. We introduced our version of GPRS and UMTS packet data access methods taxonomy which allowed us to discuss the details of private access technologies in more detail and with greater accuracy. We addressed the support of MVPNs based on CSD and packet data services offered by these systems. We also presented a brief case study, providing the reader with a high-level view of what is possible in carrier networks today and how they are likely to evolve. We repeat the same exercise in the next chapter, this time in a CDMA2000, a 3GPP2-specific environment.

# CHAPTER 7

# CDMA2000 VPN Solutions

According to Cahners In-Stat Group CDMA-based systems offering high-speed packet data services are expected to serve up to 21 percent of the worldwide subscribers by 2006. CDMA2000, just like its predecessor cdmaOne, is based on technology pioneered by Qualcomm in the United States, which then became the country with the highest concentration of CDMA subscribers and coincidentally one of the highest rates of Internet penetration. It can be reasonably expected that U.S. mobile professionals, so accustomed to ubiquitous wireline access to their corporate networks, will be more receptive to service offerings and marketing companies that advertise mobile data for business users. In fact, the response to advertisements for voice services run by Sprint PCS in the second half of 2001 that targeted business customers was extremely favorable. More recently, both Verizon Wireless and Sprint PCS have launched their 3G networks based on CDMA2000 to expand their offer with high-speed network access.

Not only in North America but also in other world regions, wireline business data users have come to expect sophisticated remote access methods accompanied by a full range of security and provisioning options. It is likely, even anticipated, that subscribers in countries such as the United States with highly developed IP infrastructures will expect a similar or better set of services to be available as a part of the next-generation wireless data offerings. We believe that for these reasons, the ability to provide secure access to private networks among other advanced data services will be of primary importance to wireless operators.

In this chapter we analyze the main types of VPN services that can be offered in the CDMA2000 system framework. We begin the chapter with an analysis of security procedures and communications between the Packet Data Serving Node (PDSN) and private networks when Mobile IP or Simple IP access methods are used. Further in the chapter we discuss various HA deployment strategies, moving on to CDMA2000 IP address assignment and AAA issues. At the end of the chapter, we present a case study outlining a real-life deployment of data services within combined CDMA2000 and legacy cdmaOne networks and exploring these systems' respective strengths and weaknesses and suitability to the task. Most of this chapter concentrates on CDMA2000 compulsory VPN methods. Voluntary VPNs, thoroughly covered in Chapter 5, are based on end-to-end secure tunneling and are generally independent of the underlying lower-level technologies. These VPNs do not vary much among different communications systems, and CDMA2000 is no exception whenever public routable IP addresses can be provisioned in the user device or when private addresses combined with appropriate address translation and IPSec NAT traversal mechanisms are available. For more on this subject, see the "CDMA2000 IP Address Management" section later in this chapter.

## Overview of CDMA2000 Private Network Access

The CDMA2000 core network data system is based on the link layer services provided by PPP combined with an elaborate multitiered mobility scheme that might optionally involve Mobile IP (as explained in Chapter 4). Therefore, compulsory VPN service offered within this system may be based on secure PPP encapsulation using one of the available mechanisms, such as L2TP, that lets a corporation perform user authentication and terminal configuration by terminating PPP sessions at an LNS it owns. Alternately, the Mobile IP protocol may be used and the PPP link layer is terminated in the carrier network; in this configuration advanced Mobile IP features for user roaming and authentication and dynamic IP address configuration are used (and in fact the development of such features in the Mobile IP protocol was driven by the CDMA2000 community). The functionality supported by one of the elements of the CDMA2000 infrastructure becomes especially important in supporting Mobile VPNs. As you might have guessed, this element is the PDSN, which handles PPP sessions originated by MS and encapsulates the user traffic for further journey through the carrier's core network or through public IP networks such as the Internet. Inversely, the PDSN terminates tunnels originated in private networks and forwards IP packets to mobile user devices or other final points of destination.

Despite the sufficient level of security available for user data traffic in CDMA2000, private network operators should be aware that compulsory tunneling lacks the end-to-end security protection that voluntary methods provide (see Chapter 5 for more on this). Whenever the decision to use compulsory tunneling is made, to ensure the desired end-to-end security level, private network operators must inquire about the security protection available for the segment of the data path unprotected by the secure compulsory tunnel, such as the radio interface and the links internal to the operator network. As with any other type of compulsory tunneling, the private network with its valuable data must trust a wireless access provider, in whose network VPN tunnels are originated or terminated. Normally, wireless operators will provide their customers with a high degree of assurance on the security level within their network, as a precondition for establishing the trust relationship necessary to run a compulsory (also known as network based) VPN service. It should be noted, however, that each operator would need to ask roaming partners to assure an equivalent level of security when roaming capability is offered as part of the service. In fact, the roaming capability will be limited in the early days of these service offerings (and also, for nationwide coverage service within the United States it will most likely never be offered, since the footprint of CDMA operators tend more and more to be national).

CDMA2000 Simple IP and Mobile IP-based VPN are no exception to the need of a trust relationship in compulsory VPN service. Though the effort was made in standards to exclude the wireless carrier from a security association between the MS and the private network, the data passed through the wireless access network is still susceptible to unauthorized access at the PDSN. Since the PDSN in wireless operator's network is both the PPP termination point and optionally the Mobile IP or L2TP origination point, the user IP packets are exposed to eavesdropping or other types of undesirable inspection within this device by persons or processes unauthorized by the private network. For this reason, the PDSN can be considered an example of a weak link in the chain of devices involved in carrying the user's traffic when used in compulsory VPN modes.

# Simple IP: A True Mobile VPN?

In Chapter 4, we reviewed the CDMA2000 three-tiered data mobility model (Figure 4.5). We mentioned that one of the mobility levels is provided by Mobile IP, preserving the mobile station's IP address when it changes its serving PDSN. When the Mobile IP service is not available for any reason, the Simple IP service is used. In Simple IP, PPP sessions originated by the mobile stations are terminated at the PDSN in a similar manner as Mobile IP. However, if the Simple IP MS must change the current serving PDSN, the PPP session is terminated and the MS must obtain a new IP address when attaching to a new serving PDSN.

CDMA2000 infrastructure vendors made significant efforts to address this problem on the data link and physical layers. One popular solution (depicted in Figure 7.8 in the "Case Study" section at the end of this chapter) is to fully mesh the network of Packet Control Functions (PCFs) and PDSNs. This solution would ensure that the MS stays anchored at the same PDSN even if the serving PCF is changing. This is possible because the PPP connection is established between the MS and the PDSN, and if the underlying network can keep the connection alive, the PPP session will be preserved. This way, the MS's IP address stays constant and can potentially even survive the MSC boundary crossings. Along with possible significant backhaul costs and IP address assignment restrictions, however, this solution will only work for the duration of the session. In other words, new dynamic IP addresses must be acquired if the session is dropped and then reinstated by the mobile or in the event of spotty coverage. As a result, Simple IP is not expected to be a mainstream access method offered to CDMA2000 VPN customers when they will require support for high mobility patterns as par of their service usage requirements. Because of these limitations, corporate subscribers using mobiles operating in Simple IP mode often can't obtain a "truly mobile" VPN service. In many instances, MSs connected to CDMA2000 networks in Simple IP mode can't maintain either their compulsory or voluntary VPN connections if the serving PDSN change happens. For these "unlucky" users, the Mobile VPN experience can be *simulated* by specially designed applications or special infrastructure enhancements (discussed in the section that follows), but never truly supported on the network layer. In addition to this hurdle, seamless roaming to the networks based on other technologies such as WLAN will be problematic at best (see Chapter 9 for more discussion). As a result, according to the taxonomy proposed in Chapter 5, the case of Simple IP should be more classified as *wireless* rather than *mobile* VPN. In summary, the Simple IP access would be optimal for access to networks requiring limited or no mobility.

## Simple IP VPN Architecture

Let's look at the architecture model of Simple IP, depicted in Figure 7.1. As in wireline remote access networks, the PPP session initiated by the MS is terminated by an NAS, whose functionality in this case is supported by the PDSN and then relayed over a tunnel to a remote tunnel termination point maintained behind a firewall in the private network. The tunneling protocol recommended by [IS835] in this case is L2TP, described in Chapter 2. The L2TP Access Concentrator (LAC) functionality supported by the PDSN, encapsulates the mobile station's PPP session and carries it over an arbitrary IP network to a remote L2TP Network Server (LNS), which terminates the PPP link in the private network.

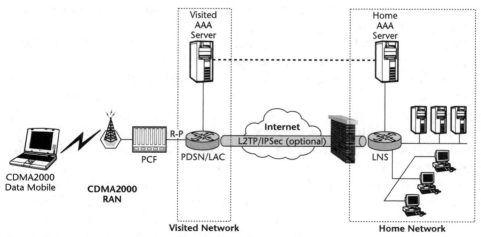

**Figure 7.1**   Simple IP VPN architecture model.

Simple IP-based wireless VPN with PDSN-originated L2TP can be classified as a typical case of compulsory tunneling. The mobile user's PPP link is effectively relayed through the PDSN over an L2TP tunnel to a remote LNS, which terminates the PPP link and, in combination with the home AAA server, provides primary authentication and address assignment functions, enabling private network owners to retain a significant amount of control over MS authentication and address assignment (and the operator, conversely, can offer the service with no need to take care of these aspects). The PDSN and its associated visited AAA server complete only enough of CHAP negotiations to discover the address of the private LNS. As opposed to Mobile IP, the Simple IP access method does not require a Home Agent but may still rely on broker-based distributed AAA infrastructure to access remote AAA servers associated with the LNS in private networks. Details on CDMA2000 AAA subsystem and IP address assignment options are provided later in the chapter. If VPN service is not requested during the Simple IP negotiation stage, the PDSN becomes an element responsible for IP address assignment and user authentication.

A Simple IP VPN protocol reference model is shown in Figure 7.2. In this figure, L2TP is augmented with optional IPSec protection. Wireless carriers often prefer this option, since as mentioned in Chapter 2, L2TP by itself is not a secure protocol and does not provide data confidentiality and data origin authentication. The L2TP tunneling is flexible, is familiar to IT departments around the world, and is often used to provide such exotic services as remote access outsourcing, IP access wholesaling, and so forth to third-party ISPs and ASP.

Simple IP-based VPN can also be used by the wireless carrier itself to allow selected users, such as maintenance personnel, access to its own private intranet. For example, operator's field technicians would be able to access essential system information or poll certain infrastructure elements status and error-reporting functions. This private intranet subnet would include dedicated LNS, which can be accessed by maintenance staff by entering, for example, their username/NAI combination (john_technicians@ support.ACME.com) to be tunneled to this LNS and then be provided access to the network service center or any other appropriate network. Because the accounting records for this access contain the NAI (Network Access Identifier) used by the wireless operator's staff, these records can be treated as "nonbillable" by the carrier billing system, or can be billed internally at a separate rate from other data services.

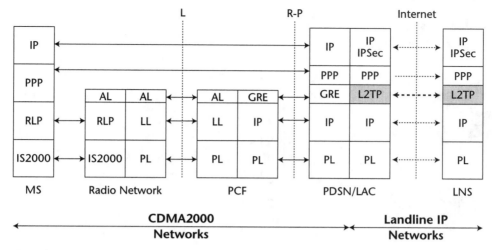

PL – Physical layer
LL – Link layer
AL – Adaptation layer

**Figure 7.2**   Simple IP VPN protocol reference model.

## Simple IP VPN Call Scenario

Let's consider the Simple IP VPN connection establishment sequence depicted in Figure 7.3. Note that this scenario assumes that a mobile station is attached to its home network—that is, the network where its original IP address was assigned.

There are two phases of VPN establishment: the establishment of connection between the MS and its serving PDSN and the establishment of L2TP session encapsulating PPP traffic between the PDSN supporting LAC functionality and an LNS in the private network. Because of the growing tendency in the vendor community to combine PDSN and LAC functionalities in a single platform and for the sake of simplicity, we refer to this combination as a PDSN/LAC in this and other chapters.

At the beginning of the first phase, the radio link is set up between the MS and BSS, and subsequent link layer connection is established between the MS and PCF (see Chapter 4 for details). To authenticate the user, the PDSN sends an authentication request to the local AAA server. The AAA server sends back an authentication response indicating whether the request

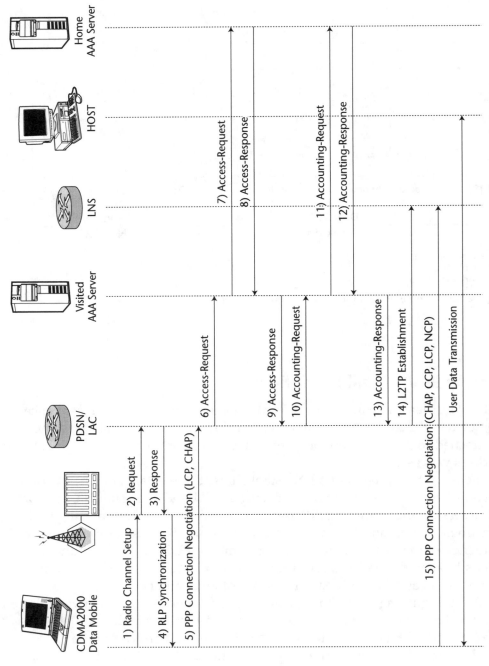

**Figure 7.3** Simple IP VPN connection establishment.

is accepted or rejected. The message from the AAA server also contains the tunnel type (L2TP in our case) and destination IP address of the LNS within the private network. If the user is positively authenticated, private network access is granted and the PPP link is established.

During the next phase, the PDSN/LAC creates an L2TP tunnel toward the LNS in the private network, if it did not exist prior to this event, with unique session created within the tunnel for individual mobile user traffic. The LNS possibly after additional LCP negotiation and user authentication assigns the IP address to the MS from the pool of IP addresses owned by the private network using RADIUS, DHCP, or another dynamic address assignment mechanism at the NCP establishment phase. Next, the LNS strips off the encapsulation headers and routes the IP packet to the destination host within its private network. In the other direction, IP packets from a host trying to send packets to an MS, arrive at the LNS, which encapsulates them in PPP frames and send them to the PDSN/LAC where the MS is anchored through the L2TP tunnel. The PDSN/LAC strips off the L2TP header and forwards the PPP frames to the MS. As mentioned previously, IPSec can augment this scenario, by securing L2TP tunnels with ESP transport mode. If the MS changes its location and attaches to another PDSN, this whole procedure must be repeated and new IP addresses may be assigned, which may inconvenience many MVPN service subscribers.

## Mobile IP-Based VPN

CDMA2000 Mobile IP VPN service, standardized by TIA and 3GPP2, addresses many of the shortcomings of a Simple IP-based solution by preserving MS IP addresses while it migrates through the area covered by more than one PDSN. For this reason, Mobile IP VPN can be classified as a truly mobile service according to the taxonomy from Chapter 5. Mobile IP VPN can be implemented in two basic ways in CDMA2000-based systems, as shown in Figure 7.4. The first one, which we'll refer to as *remote public HA VPN*, assumes that the HA resides in a private network other than wireless carrier's and will be connected with a PDSN, which resides in wireless carrier's domain via a secure Mobile IP tunnel. The other one, which we'll refer to as *local private HA VPN*, assumes that the HA will reside in the same intranet as a PDSN and will be owned and maintained by a wireless carrier. The VPN services in this case will be supported by a combination of Mobile IP tunnels and arbitrary transport options, such as chaining with other tunnels, leased lines, or ATM PVCs. In the following sections we discuss both options.

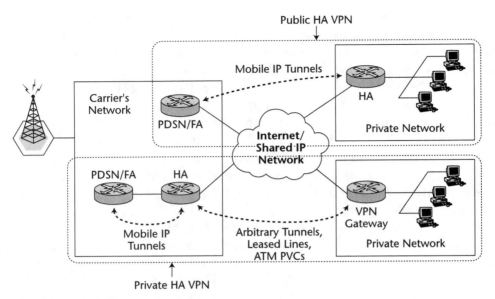

**Figure 7.4**  Mobile IP VPN options.

## Public HA VPN Option

This Mobile IP VPN option is well described in IETF, 3GPP2, and TIA/EIA standard documentation.[1] It is expected to be a mainstream scenario supported by the majority of CDMA2000 wireless operators offering MVPN services. In this approach, depicted in Figure 7.5, all of the downstream traffic (to the MS) that originated in a private network is tunneled from the HA that resides there to the PDSN that resides in the carrier's network. The upstream traffic (originated by the MS) is tunneled from the PDSN in the wireless access network to the HA in the customer network. For this purpose, the PDSN may establish an optional reverse tunnel, as defined in [RFC3220]. Both forward and reverse Mobile IP tunnels are based on IP in IP or GRE protocols and can be combined with optional IPSec security.

The IP address of the MS is assigned from the address space of the customer network, which might rely on public or private IP addressing schemes, relieving the wireless access provider from the duty of IP address management, similar to a Simple IP VPN case. As defined in [IS835], the address of the HA in the private network is discovered using NAI in Mobile IP RRQ when HA is statically allocated (dynamic HA allocation is

---

[1]In fact, this is the only Mobile IP mode VPN option defined in the standards, which is why we needed to create a specific name for it to help readers distinguish it from the other option, defined by the carrier community.

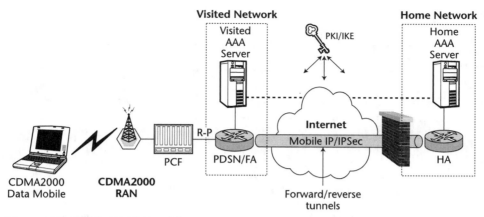

**Figure 7.5**   Public HA VPN architecture.

considered in the "CDMA2000 IP Address Management" section later in this chapter). The MS in this scenario must register with both the visited and home AAA servers and undergo a AAA procedure involving AAA clients in both the PDSN and the HA.

### Public HA VPN Security

Mobile IP tunnels to and from private networks established through public IP networks such as the Internet are generally insecure and require security protection, similar to the situation with the L2TP tunnels in the Simple IP case. Such protection can be provided by the IP Security protocol suite combined with a mechanism for distributing keys, like the Internet Key Exchange (IKE) discussed in Chapter 2. Figure 7.6, depicts a protocol reference model for this VPN option. It is important for the HA to verify the identity of the wireless carrier's PDSN that will have access to unprotected user data for the duration of the session. It is also important for the PDSN to verify the identity of the HA so that user traffic is not misdirected to an unknown insecure location. In the public HA VPN access scenario, the HA is owned and operated by the private network to which the user is gaining access and will manage both security and mobility of the user by forming dynamic security associations with changing serving PDSNs.

**NOTE** Some of the mechanisms we describe in this section are unique to CDMA2000 and are not a part of the generic 3220 Mobile IP spec. Apendixes A and B list Mobile IP extensions and RADIUS attributes that have been added to the original IETF specification by 3GPP2 and TIA.

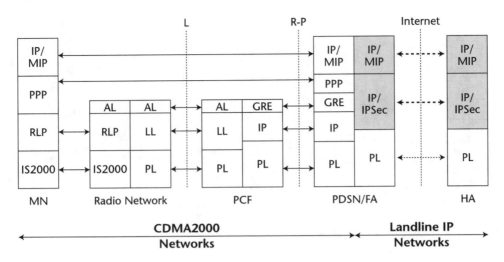

**Figure 7.6** Public HA VPN protocol stack.

Normally, wireless carriers deploy IP security for interdomain communication and for Mobile IP signaling protection. To further minimize the chances of a security violation, the traffic to and from the HA can be encrypted by using one of the techniques outlined in Chapter 2. The PDSN can decide which policy to apply based also on the *Security Level* RADIUS attribute as defined in [IS835]. During a setup of an IPSec-secured tunnel between the PDSN and HA, IKE is used to verify the identity of the PDSN and HA. The security key association may be:

- A statically configured secret for the Mobile IP HA-FA authentication extension
- A statically configured IKE pre-shared secret
- A dynamic IKE pre-shared secret distributed by home AAA
- PKI with certificates

The *static Mobile IP HA-FA authentication extension* supersedes the static IKE pre-shared secret, which supersedes the dynamically distributed IKE secret, which supersedes the PKI certificate in order of precedence. Preprovisioning the MN-HA pre-shared key is required in the current version of

[IS835], the standard governing most of the requirements of the CDMA2000 core infrastructure. Keying material that is distributed during the AAA registration process should be protected against eavesdropping. If an attacker could learn these keys, he or she could carry out denial-of-service or session stealing attacks against mobile nodes. This protection can be provided on a hop-by-hop basis—for instance, by using IPSec between visited AAA servers in the rest of the AAA infrastructure.

When a *static pre-shared* secret key association is used, the first phase exchange is authenticated with message authentication codes. Using pre-shared secrets can simplify operation because it avoids the need to process and validate certificates. However, these associations may introduce an additional burden because of the need to establish them in PDSN-HA pairs. For this reason, [IS835] provides a mechanism for *dynamic pre-shared keys distribution* via the AAA infrastructure during mobile node registration. While the home AAA processes and validates the MN's challenge-response, it can generate a pre-shared secret and distribute it with the AAA response back to the PDSN. The PDSN then can use the secret, along with an identity constructed from the response, to carry out IKE negotiation with the HA. This enables secure establishment of IP security between the PDSN and HA without the need for manual configuration of the keys between all possible pairs.

If reverse tunneling is supported by the Home Agent as indicated by the home AAA server in the [IS835] *Reverse Tunnel Specification* RADIUS attribute (see Appendix B), IPSec security is authorized for tunneled data, and the mobile requests reverse tunneling, then the PDSN will provide security on the reverse tunnel. Reverse tunnels are set up as a result of the mobile node setting the "T" bit in its registration request. This causes all packets from the mobile node to be encapsulated and delivered to the HA by the PDSN. Such tunnels enable the use of private, nonunique addresses by mobile nodes, and the reverse (as well as forward) tunnels can be optionally protected with IPSec if desired by the home domain.

## Private HA VPN

Wireless carriers might not always be open to the idea of sharing their data infrastructure with the rest of the world like in the public HA VPN model. This might be especially disconcerting when some of the key infrastructure elements such as HA owned by a third party will be connected to their core networks through the shared public Internet. Also, carriers might not be willing to give up control of their subscriber management and hesitate to become "just" wireless data access providers.[2] These concerns were significant to GPRS operators and standard bodies like 3GPP and eventually prompted them to recommend placing all of the GPRS infrastructure elements in the wireless operator's domain (as explained in Chapter 4). Similar to their GPRS and UMTS counterparts, CDMA2000 operators deploying the *private HA VPN* option will own both PDSN and HA infrastructure elements.

While significant amounts of HA capacity must be provisioned in the wireless carrier's network for non-VPN use to enable them to offer ISP and other services, use of the carrier's HAs for VPN services have not been addressed by the standards and therefore must be analyzed carefully. The CDMA2000 subscriber's data path will always include both the PDSN and the HA, at least in one direction. The downstream data traffic must pass through the HA in the MS home network and the serving PDSN. The upstream traffic (from the MS) must pass through the PDSN only if the MS requests plain Internet access, and through the PDSN/HA pair joined by the reverse Mobile IP tunnel if the MS requests private network access. To satisfy these requirements, wireless carriers must deploy enough HA capacity to support Mobile IP MSs whether or not they require private network access or only requesting plain Internet access.

Having such a vast HA infrastructure already available, wireless carriers wishing to exercise as much control as possible of the subscriber provisioning might then completely disallow the access to the HAs in the private networks, forcing all the traffic to and from private networks through their own HAs and then forwarding it between private networks via the use of other technology, as shown in Figure 7.7. In this scenario, private networks do not need to maintain the HA and terminate Mobile IP tunnels. Instead, the wireless carrier and the private network must rely on a set of tunnels (or other technologies) concatenated at the carrier-owned HA combined with private peering agreements and specific SLAs to be able to provide secure VPN offerings.

The rules of deployment for private HA VPN are quite different from those for public HA VPN and bring some significant consequences for operators. Arguably, private HA VPN might simplify IP address assignment for mobile users by allowing one entity, a wireless carrier, to control the procedure. Optionally carriers may even be able to consolidate—at least in

---

[2]Data networking industry jargon for such operators in the wireline world is "dumb pipe providers." From this telling title it is obvious why many wireless operators are trying to avoid such status.

theory—the IP address assignment process into one location, a hypothetical HA farm combined with massive IP address repository and super-sized DHCP and AAA servers. Moreover, wireless carriers can retain control over the user's provisioning and both payload and signaling traffic security, which will minimize the risks of their core network security breach. However attractive this reasoning might be for both greenfield and established operators, in our view the disadvantages of this solution equal if not outweigh its advantages. To name a few, operators favoring private HA VPN will have to pick up the burden of user provisioning, fully engage in ISP-style business and hence competition, be responsible for supplying millions of IP addresses to their mobile users, and be prepared to limit their offerings to corporations and forgo other advantages and simplicity of standard-based public HA VPN.

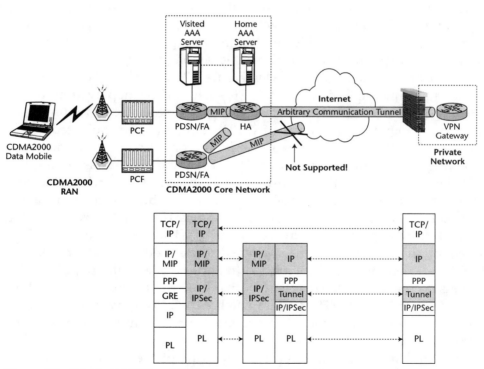

**Figure 7.7** Private HA VPN architecture and protocol stack.

The responsibility for IP address assignments might pose a dilemma for a successful[3] CDMA2000 operator. The operators must decide whether to provide their subscribers with a public or private "topologically incorrect" IP address or a mixture of both. Both cases would be equally problematic. Public routable IPv4 addresses are a scarce commodity and unlikely to be available in the millions. Private IP addressing might seem to be an easy way out, but that practice would prevent mobile subscribers from accessing private networks using voluntary VPN based on end-to-end tunneling, which requires public routable IP addresses, unless complex and not adequately defined NAT traversal schemes are deployed within the operator's core network. In any case, customers will effectively feel forced into using private HA VPNs, often entailing mandatory customer-operator agreements as the only option available for private intranet access.

Another potential challenge associated with private HA VPN is the necessity to create a tunnel-switched infrastructure around the HA. This situation is not quite addressed in standards and will require a new architectural framework involving wireless carriers and their corporate peers alike. Creation of such a framework will be a nontrivial exercise involving new types of SLA definitions, new billing arrangements, and new requirements to HA platforms—which will need to support tunnel switching and WAN technologies on a carrier-class scale, among other new tasks.

Sample architecture, as depicted in Figure 7.7, can be based on one of the tunneling technologies defined by IETF, such as IPSec deployed in tunnel mode or combined with MPLS. In such a scenario the Mobile IP tunnels to and from geographically distributed PDSNs would be terminated at the private HA in the carrier's network and then concatenated with the corresponding IPSec tunnels created for individual enterprises, assuming the existence of predefined relationships with the wireless carrier. This scenario assumes that not only IP address assignment but also authentication of mobile stations will be completed in the carrier's network. A similar "tunnel switching" scenario in GPRS and UMTS, referred to as the *PPP Relay* option under the *Nontransparent* IP access method, is *well defined* by 3GPP and therefore, can be expected to become one of the popular VPN options, as mentioned in Chapter 6. A likely alternative might be based on use of private lines or ATM or Frame Relay PVCs to transmit user data to and from corporate network. The details of such a solution, however, are beyond the scope of this book.

We complete this section by discussing one of the most controversial aspects of the private HA VPN solution—one that does not concern the technology itself but rather the technical community's perception of it. The

---

[3]That is, the one with the number of subscribers in the millions.

standards bodies such as the IETF Mobile IP Working Group (which defines Mobile IP and AAA systems architecture) and TIA [TR45.6] (which defines CDMA2000 packet data core) based their efforts on the assumption of access network transparency and relative subscriber independence from the access providers. Their efforts concentrated on definition of Simple IP or Mobile IP public HA-based private network access methods, both of which preserved a certain degree of enterprise control over mobile user provisioning.

It was assumed that private networks wishing to provide their members with mobile access would be able to achieve this goal through a relatively straightforward standards-based set of agreements with wireless carriers that would not require them to sacrifice their involvement in AAA, IP address assignment, and other remote user management functions. As such, while not directly violating the CDMA2000 private network access standards and overall logic, the private HA VPN approach works against at least the spirit of the standards This controversy is bound to stir emotions and spark debates in both wireless and corporate IS communities with consequences yet to be measured. After all, the original goal of Mobile IP was focused on providing easy secure private network access in the mobile environment, ease of distributed infrastructures implementation, including AAA brokerage, and support for a wide range of options to control remote user provisioning exercised by service providers and private networks alike.

# HA Allocation in the Network

In this section we discuss HA deployment options in CDMA2000 core networking as well as its impact on MVPN offerings and architecture. We start with an analysis of HA allocation in the network relative to the PDSN, then continue with a description of dynamic HA allocation option. Finally, we discuss the importance of a fault-tolerant HA in the CDMA2000 core network.

## Private HA Allocation Relative to the PDSN

The problem of HA allocation in the operator's network is somewhat related to dynamic HA allocation addressed later in the chapter. You'll recall from Chapter 4 that by definition a PDSN must cover a certain geographical region, while an HA, representing MS home network, serves as an anchor point for the data session. PDSNs serve both homing and roaming users who are currently in a particular network or region, while the HA always serves the same set of provisioned users regardless of whether those users are attached to their home network or roaming far away from it. In this respect, there are two main HA allocation scenarios to be considered: collocated HA and centrally located HA.

### Collocated PDSN/HA

In the *collocated HA* allocation scenario, there are more than one HA locations in the network. Since all Mobile IP user traffic (at least on the uplink) must pass through PDSN/HA pairs, the quantities of PDSN and HA ports provisioned in the systems should be very close—especially if the private HA VPN access method is implemented. Often these functionalities will be supported in the same or similar platforms, so it makes for good economies of scale to collocate, or cluster, them in some selected geographical locations, like regional data centers.

The main advantage of this approach lies in the ability to dynamically reprovision local PDSN/HA clusters if the customer mix—that is, the roaming versus homing mobile ratio—changes. For example, during a trade show or any other professional event gathering large groups of users with mobiles assigned to HAs serving other geographical locations, more mobile users than usual will have to be served by local PDSNs, which would have to tunnel traffic to HAs around the world. To address this situation, carriers deploying collocated HAs will be able to easily reprovision local PDSN/HA clusters for more PDSN capacity. When the event is over, the cluster can be changed back to the usual ratios.

Another advantage of this approach is for the carriers expecting to serve large numbers of stationary users in different locales of the country, such as wireless carriers competing with local telcos for the wireline local and long-distance phone markets. Since there is not much mobility in such networks, the users usually stay within the regions served by local HAs, allowing operators to minimize the use of their backhaul network. Similar levels of backhaul optimization will be achieved for the networks with the majority of highly mobile roaming users when dynamic HA allocation is available, thereby sidestepping the problem of triangular routing.[4]

Finally, when collocated HAs are used, each PDSN/HA cluster can more efficiently utilize its address management capabilities by being able to allocate IP addresses to the MSs from the pools of IP addresses provisioned locally. While some inefficiency may result because of the disjoint IP

---

[4]Triangular routing happens when packets have to (potentially) traverse the wide-area Internet twice, once on their way from correspondent node to HA and again from the HA to the PDSN.

address pools, the size and scale of the PDSN/HAs should be sufficient to guarantee good average-case utilization. Private addresses and NAT may also help alleviate concerns about address space.

### Centrally Located HA

In the *centrally located HA* scenario, the HAs serving all Mobile IP users in the network are located in a single data center. This solution bears some advantages (in the absence of dynamic HA allocation, see the section "Dynamic HA Allocation" coming up in the chapter), especially for the operators serving the users, a majority of whom are highly mobile and often change their PDSNs, and must therefore be terminated back at their original HA. Single HA data centers provide greater ease of management, including provisioning, maintenance, and upgrade, for operators. In addition, since spare resources and backups are shared better at a centralized location than at distributed sites, disaster recovery is potentially easier than with a collocated HA solution. Another advantage is the greater possibility to provide HA load balancing that includes all of the HA's capacity in the network, as opposed to small-scale load balancing within only the local HA cluster in the collocated HA model.

The centrally located HA option might appeal to those operators wanting a centralized location to hold and manage a pool of IP addresses for networkwide assignment to mobile users to utilize them more efficiently.

#### HA Reliability

HA reliability becomes especially important in the centrally located HA model, and therefore its implications must be considered in a separate subsection. An MS can be served by any available local PDSN. In case of PDSN failure, the MS behaves similarly to the event of PDSN relocation by sending messages soliciting advertisements until a standby PDSN comes into service. Both voluntary and compulsory tunnels are not be affected by this event—provided that the inactivity timer and other parameters of the MS are properly configured. Therefore, the PDSN failure may not be a catastrophic event and may be gracefully resolved, thanks to the properties of Mobile IP.

The effects of HA failure on the MS—in both public and private HA VPN scenarios—are more profound and can have devastating consequences for MS data connectivity. In CDMA2000, each Mobile IP MS is programmed to access only one specific HA. This means that if the HA provisioned with the IP addresses of a certain MS group failed, all of the MSs associated with this HA will not be able to receive packet data service. To remedy this situation, the HA platform must therefore include both extensive internal and inter-chassis failover options, which would, for example, automatically associate the IP addresses provisioned in the failed HA to another hardware element within the local HA cluster.

This high-level overview of HA allocation options and reliability shows that there is no compelling reason for a carrier to adopt one model over the other, since both models have their fair share of advantages and constraints. We believe that real-world private HA deployment models should include both, which would provide CDMA2000 operators with a variety of options and allow them to flexibly and dynamically allocate their core networking resources as business conditions change.

## Dynamic HA Allocation

Our previous discussion was based on the assumption that the Home Agent in the CDMA2000 core network can only be provisioned statically. This was done mainly because of the current status of dynamic HA allocation standardization, which as of today is not completed. Standard groups such as IETF, 3GPP2, and TIA are currently working on extending the CDMA2000 core network standards framework beyond those in the current IETF RFCs by adding support for dynamic configuration of a mobile node's home address (see Appendix A) or the HA itself.

In the current architecture the MS is hard-coded with the address of a particular HA, which is included it in its registration request during the PDSN signup procedure. A static HA is simpler to support, because the HA IP address is configured into the mobile node and a static shared secret can be used for the MN-HA authentication extension. However, a dynamically assigned HA, collocated or located near a PDSN, can provide for significant optimization of operation because of greater service availability and more optimal routes in the case where a MS is roaming far enough from its home network to make backhaul expenses substantial (for example, data from a PDSN in Alaska would not have to be shipped to and from an HA in Texas every time the user wants to read a few emails from a mail server located in Seattle, if the HA was possibly dynamically assigned to a home agent nearby). In addition, dynamically assigned HAs address the problem of triangular routing, mentioned earlier.

While certainly desirable, this feature requires a complex security arrangement, which is why the standardization of this option is taking such a long time. Let's look at what is involved in supporting secure dynamic HA allocation in a CDMA2000 core network. Figure 7.8, based on current standards drafts, details the steps necessary to dynamically allocate an HA.

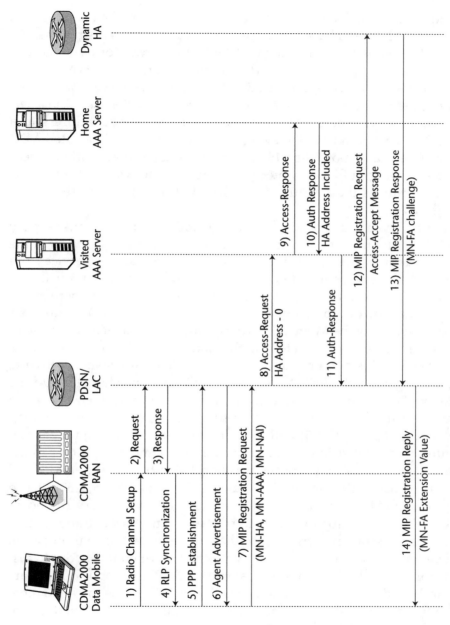

**Figure 7.8**  Dynamic HA allocation establishment.

Dynamic HA establishment requires a development of a shared secret between the MS and HA so that subsequent mobility registrations can be authenticated as the MS changes PDSNs. In the case of dynamic HA allocation, the address of the HA is determined by an AAA server and not by the MIP RRQ (Mobile IP Registration Request), as is the case with static HA assignment. A home AAA server dynamically allocates an HA in a service provider's or remote private network and returns its address to a visited AAA server and PDSN, along with dynamically distributed MN-HA shared secrets to both the MS and HA for later authentication. These secrets are cryptographically protected by the AAA network in transit. The PDSN then returns these values to the MS, which begins to use its new home address.

To support dynamic allocation of a home address, the MS must supply an NAI in its Mobile IP Registration Request. This is a unique name of the form user@domain that identifies the user who is requesting service from the network. It acts as an identifier and is not associated with the IP address of the underlying device. The NAI enables the serving network to find the home network (possibly located in a private network) via an AAA infrastructure. Using the Mobile IP *Challenge/Response* extensions, the user's credentials may be authenticated by the home domain. Once the user is authenticated and is authorized to receive service on the visited network, the MS can register with the HA. Because an NAI and not a home IP address appears in the registration request, the home agent may then allocate a home address for the mobile node and return it in the registration response.

The next release of [IS835] will include a dynamic HA assignment feature with dynamic distribution of keys from the home AAA server to the HA. This assumes that HAs are always allocated in the home network and have security associations with the home AAA server. [IS835] C3 will also define a new RADIUS-based mechanism for the HA to query the home RADIUS AAA server for the key, after it has been allocated and after it receives the registration request from the mobile node. Normal operation would be for the mobile to deregister with the HA when it is about to disappear from the CDMA2000 packet data network. If the mobile disappears temporarily and reappears at another PDSN, the MS will be forced to renegotiate PPP and reregister with the same HA. If this reregistration does not happen, the Mobile IP binding will exist on the HA until the Mobile IP lifetime expires and the HA resources are freed.

## CDMA2000 IP Address Management

In this section we provide an overview of IP address management from both a wireless carrier and a private network perspective. When an MS connects to a private network in either Simple IP or Mobile IP modes, it may be assigned a private IP address out of private network address space. Because no global authority allocates such addresses, they may not be globally routable or even unique, which should not pose a significant problem to the enterprise or wireless carrier. It is important, however, to architect the PDSN so that it can properly handle such a situation, that is, the PDSN must be able to appropriately route packets to and from HA's even though they have overlapping private addresses. To accomplish this, the PDSN must make use of the HA address in the outer IP header of tunneled packets and the link layer identification information on the access network side (that is on the R-P interface side) of the PDSN to resolve potential collisions in the IP addresses assigned to different MSs.

While private addresses are perfectly acceptable in the CDMA2000 VPN environment, public addresses allowing for easy voluntary MVPN might mean additional benefits to corporate CDMA2000 service subscribers. For example, there might be a requirement to provide end-to-end security for the protection of such extravagant data types as classified information, in addition to different levels of security provided by wireless carrier to different sets of customers. In such cases, the IT department might need to support voluntary VPN based on end-to-end tunneling such as IPSec or similar techniques to add an additional level of protection for sensitive data and to limit the exposure of private data to a third party such as a wireless access provider.

Alternatively, for the wireless carriers relying on private IP address space in their core networks and using network address translation to maximize the efficiency in dealing with scarce public IP addresses, voluntary MVPN can also be supported (regrettably, with greater difficulty) when one of the available NAT traversal mechanisms is properly executed by the carrier. (See Chapters 2 and 5 for more on these issues.)

## Simple IP VPN Address Assignment

In CDMA2000, Simple IP address assignment is handled by the PDSN unless VPN service is requested. As opposed to Mobile IP, the Simple IP access method does not allow for static addresses to be preprovisioned in the MS. Instead, the IP address *must* be assigned to the MS dynamically via one of the available address assignment mechanisms, during PPP startup when the MS first registers with the PDSN and sends an IP address 0.0.0.0 during the IPCP phase to request a dynamic IP address. Note that the address assigned to the MS may be a private address as per [RFC1918] or a public address.

The following list outlines the IP address assignment options available for Simple IP:

- Assignment from a pool of addresses configured in the PDSN or in a PDSN cluster. The pool may be statically associated to the user via a mapping table provisioned on each PDSN, or the name of the address pool may be returned to the PDSN in the RADIUS Access Accept message by the AAA server.

- Assignment via the use of an AAA server such as RADIUS or DIAMETER when performing authentication of the MS. Like the local pool case, the address from the AAA server is communicated to the client during PPP negotiation.

- Assignment via DHCP, which requires DHCP client support in the PDSN.

When compulsory VPN service is requested in Simple IP mode, the responsibility for IP address assignment to the mobile is transferred to a private network. As described in the "Simple IP: A True Mobile VPN?" section earlier in the chapter, in this case the PPP link is terminated and then encapsulated into L2TP tunnel and forwarded to the LNS in a private network, where address assignment then occurs.

## Mobile IP VPN Address Assignment

Like the Simple IP service, the address assignment process for Mobile IP service can be provided via a variety of options. Unlike Simple IP, however, mobile stations requesting Mobile IP service can optionally be provisioned with static preconfigured IP addresses. When an IP address is statically assigned to the MS, it will be proposed to the PDSN via IPCP during PPP negotiation. Recall from Chapter 4 that the IP address assignment for Mobile IP service in CDMA2000, and for Mobile IP in general, is always handled by the Home Agent. This makes the HA, in both public and private form, the most important element in IP address assignment process in the Mobile IP VPN service.

After the MS is authenticated with the PDSN, it can request either static or dynamic IP addresses from its HA. The HA returns the IP address to be used by the MS in the Mobile IP Registration Reply message, which is forwarded to the MS by the PDSN. As outlined previously, this address may be either publicly routable or provided from the *private* address space at the discretion of the wireless carrier (in the case of the private HA VPN option) or a private network (in the case of the public HA VPN option).

Multiple, overlapping private addresses are supported by the PDSN per the TIA [IS835] standard, as long as the addresses from each *individual* HAs are unique and nonoverlapping. Another useful option that further distinguishes Mobile IP address assignment capabilities from Simple IP is the ability to support multiple IP addresses in the MS to support multiple communications sessions between the MS and its private network (somewhat similar to the concept of multiple PDP contexts support in GPRS and UMTS networking).

If the MS requires an access to a private home address, then it has to negotiate reverse tunneling as described in [RFC2344]. As a result, the PDSN forms a logical association that contains the R-P Session ID, the mobile station's home address, and the Home Agent address. When the PDSN receives a packet for a registered mobile station from the HA, the PDSN maps the mobile station's HA address and the home address to one association, and transmits the packet on the R-P connection indicated by the R-P Session ID of the association.

# Authentication, Authorization, and Accounting for MVPN Service

Neither Mobile IP nor L2TP by themselves provide scalable mechanisms for access control or accounting. Basic Mobile IP does specify extensions that can be used to authenticate the MN to the FA or the FA to the HA, but these extensions are not mandatory and they assume the existence of preconfigured shared secrets between these entities. That creates a problem because the worldwide public CDMA2000 cellular network will include many subnetworks or domains owned by a variety of private enterprises, CDMA carriers, wireline ISPs, and ASPs. Visited wireless carrier's networks that support PDSNs will expect payment for wireless data services from the mobile user or the user's home domain. To obtain assurances of payment, the CDMA2000 core architecture must support scalable AAA networks consisting of interconnected AAA servers that offer an array of services, as opposed to a group of disconnected noncommunicating AAA servers.

## CDMA2000 AAA Architecture

In Chapter 4, we described the basic concepts of AAA in the CDMA2000 environment, which are largely based on the use of RADIUS and other protocols such as PAP and CHAP. In this section we analyze in more detail the

CDMA2000-style AAA architecture and its impact on MVPN services. To provide a robust AAA functionality for private network access, the concept of a distributed visited-home AAA infrastructure had to be taken one step further. To better satisfy requirements for different methods of private network access and facilitates peering without the need to preestablish agreements, it evolved into the form of *visited–broker–home* AAA architecture, shown in Figure 7.9. This architecture was developed with a goal somewhat similar to GRX in GPRS networking, discussed in Chapter 6, to facilitate a network architecture that can be shared by many private peering entities, including ISPs, ASPs, corporate networks, and wireless carriers. Note that the need of GRX in GPRS is for BGP peering, rather than AAA servers peering.

The MS accessing a *private* network through the *access* network provided by a third-party entity must be authenticated by both networks. As a result, the MS is identified to the access network by its ID, such as an International Mobile Station Identifier (IMSI), and to a private network via an NAI. As shown in Figure 7.9, such authentication requires AAA functionality at both visited and home networks. In CDMA2000, this functionality is achieved through a visited network RADIUS AAA client (usually hosted by the PDSN) and server, and home network RADIUS AAA client (usually hosted by an HA for Mobile IP and an LNS for Simple IP VPN) and server.

In addition to authenticating and authorizing MS in generic CDMA2000 communications, when private network access is requested, authentication requests are forwarded from a *visited* AAA server associated with the PDSN to a *home* AAA server associated with the HA, and authorization responses are forwarded in the other direction. The accounting information must then be stored by the visited AAA server and optionally sent to the home AAA using a reliable AAA protocol and then forwarded to a billing system. For VPN service, the accounting information may include parameters such as NAI, QoS, Session ID for Simple IP service, and destination address. The AAA home-visited communication is based on a reliable transport mechanism, can be optionally protected by IPSec, and is capable of distributing a pre-shared secret for IKE.

This model is fundamentally similar for both Simple IP and Mobile IP VPN types and might include optional components such as a RADIUS proxy server and AAA brokers (described in the section that follows). For both CDMA2000 VPN types, basic communication between RADIUS clients and servers is governed by [RFC2865] and [RFC2866], details of which are beyond the scope of this book. This communication can also be optionally secured by IPSec to provide a security association between the MS, the PDSN, and the HA (or LNS in the Simple IP case) and support dynamic key distribution using IKE.

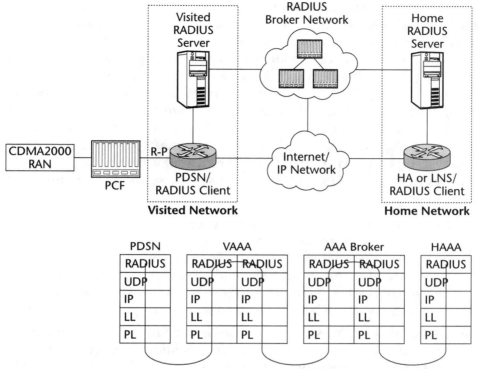

**Figure 7.9**   CDMA2000 RADIUS-based AAA architecture and protocol reference model.

## CDMA2000 AAA Brokerage

The home/visited AAA infrastructure just described was designed to serve home and visited network with the relationship preestablished via a set of SLAs. In cases when such relationships have not been established but the visiting MS requests data service, the AAA *brokerage* must be used. Since this statement, while being accurate, provides little in the way of explanation, we need to take a closer look at both the problem and solution. Home and visited AAA servers may have a direct bilateral relationship. The cellular architecture, however, will involve thousands of domains with many private networks owned by enterprises that seek wireless data service for their mobile workforce. If the number of domains were small, the serving and home networks could have preexisting relationships (perhaps secured via IP security associations). However, this would not be a scalable solution; it would require too many pair-wise relationships to be pre-established.

AAA brokers host directories that allow AAA requests to be forwarded based on the NAI to home networks or to other brokers knowing the whereabouts of the home network. Brokers may also take on a financial role in the settlement of accounts between domains and may process accounting records for the network access requests they authorize. Because the visited network will not provide service unless it is able to obtain authorization from the mobile user's home network, or from a broker that accepts financial responsibility, the AAA infrastructure must be reliable. This implies that servers must retransmit requests and switch over to backup servers when a failure of the primary unit is detected.

As defined in [TSB115], AAA networks must support three modes of broker operation:

- Nontransparent (nonproxy) mode, where the brokers essentially terminate requests to and from visited and home AAA servers and initiate new requests on their behalf. In this mode the broker is allowed to change the content and attributes of the messages. This mode is most likely to be used when the broker is permitted to financially act on behalf of visited networks.

- Transparent mode, where the broker is not authorized to make any changes to the AAA messages and is only allowed to redirect them to appropriate points of destination.

- Redirection mode, where broker AAA server refers the service provider to another AAA server.

Another important task of an AAA broker is to facilitate roaming services. Roaming services permit mobile users outside their carrier's area of coverage to use other carriers' networks to access the public Internet or private networks.

## Mobile IP VPN Perspective

In the case of Mobile IP VPN, when an MS is accessing an HA in the private network, the wireless access provider (that is, the PDSN owner) must not be involved in the security association between the MS and its home network. That was another requirement that the CDMA2000 AAA architecture needed to comply with. With the TIA Security Level attribute in the Access-Accept message, the home AAA server is able to authorize the PDSN on a per-user basis to optionally use IPSec on the registration messages and the tunneled data.

If the home AAA server has indicated that an IP security association must be used between the PDSN and HA, the PDSN will provide IPSec services as indicated in the [IS835] based on a 3GPP2-defined security level attribute (see Appendix B). If no security association is in place, the PDSN attempts to establish the security association using the HA X.509 certificate. If no HA X.509 certificate exists, but the root certificate exists, the PDSN attempts to establish the security association using X.509 certificates it received in Phase 1 IKE. If the certificates do not exist, the PDSN attempts to use the dynamically distributed shared secret for IKE received in the Access-Accept message. If no shared secret was sent, the PDSN attempts to use a statically configured IKE pre-shared secret, if one exists. If the PDSN does not receive the 3GPP2 security level attribute from the home RADIUS server, and an IPSec security association to the HA already exists, the PDSN continues to use the same security association. If no security association exists, then the PDSN follows locally configured security policy.

## Simple IP VPN Perspective

For Simple IP VPN access mode, the AAA architecture does not include the HA. Its functionality is instead supported by the LNS, as described earlier in this chapter. The AAA server must locate the LNS providing access to the user's home network. Proxy AAA servers in related service providers' networks must therefore be set up and provisioned in order to locate the LNS. After the LNS is located, an L2TP tunnel is established between the LNS and the PDSN from which the MS requests services. The MS IP address is assigned by the LNS after the tunnel establishment. Since the MS does not participate in making routing decisions between the tunnel endpoints, both registered or unregistered IP addresses can be assigned to the mobile node. The visited AAA server records the accounting record and forwards the Accounting Request message to the home AAA server if roaming is involved.

Let's consider the sequence of AAA events taking place during Simple IP private network access initiation. When a Simple IP MS initiates the connection to the PDSN, the PDSN forms an Access Request message and sends it to the home AAA server for authentication. The request is positively authenticated and an Access Accept containing tunneling type "L2TP" is sent back to a visited AAA. The PDSN initiates the L2TP tunnel if it has not been established prior to this event, and it sends an Accounting Request message for billing purposes to record the starting time of the service. When the service is no longer required, the L2TP user session and tunnel are terminated. Finally, the PDSN sends another Accounting Request message to record the time when service stopped.

## Case Study

The CDMA market historically developed under different rules than GSM, and CDMA2000 is expected to carry an even greater level of dissimilarities from UMTS despite their radio interfaces are closely related. CDMA2000 coverage is expected to continue to be concentrated around two large clusters in North America and East Asia (China, Japan, the Pacific Rim, and Korea), with deployments in Latin America, the Caribbean, and Eastern Europe possibly seeing negative growth. To date, the CDMA arena has been dominated by a few very large carriers, with smaller ones focused on serving niche markets and rural areas. In this case study, we analyze a CDMA2000 network (combined with the existing IS-95 network) built in the United States by a large carrier that recently merged with ACME Wireless, who, as described in the case study in the previous chapter, constructed a network in Europe. This newly established North American arm of ACME Wireless was recently renamed ACME USA.

This operator is now well funded and striving to provide its North American customers with an array of services similar or even surpassing that offered by its partners in Europe. Given the higher efficiency of the CDMA2000 air interface compared with TDM-based GSM GPRS, this just might be possible, and sufficient bandwidth should be available to provide mobile data users with advanced services. While GPRS operators, especially in urban centers like London, are already struggling to provide their subscribers with more than one time slot, this problem should be easier to address with CDMA2000's high-efficiency spread spectrum radio network.

ACME USA has been offering circuit data service for a few years now. Its core data network was based on industry-standard Interworking Functions (IWFs) concentrated in six data centers evenly covering the geography of the United States and hosting application servers and other equipment. While the network had been rolled out nationwide from the beginning, the take rates have not met expectations, which resulted in less profit but also in less data networking maintenance and support overhead, as well as in lower than originally planned rates of expansion. As a result all IWFs had been concentrated in one location to reduce maintenance costs. One of the few new features was a "Quick VPN" service marketed to business users based on Quick Net Connect architecture offered by the IWF manufacturer, allowing for fast private network access using L2TP tunneling to a private network instead of a traditional modem dial-up procedure.

Last year, ACME USA decided to upgrade their network to CDMA2000 by initially rolling out nationwide pilot packet data service. Under this plan, the existing data centers would be utilized to host PDSNs and HAs farms for plain Internet service, as shown in Figure 7.10. For this upgrade to be successful the radio equipment requires software and hardware upgrades, MSCs must be augmented with the PCFs implemented on a separate platform, and the network of PDSNs and HAs must be installed. ACME USA's engineering and marketing staff performed an extensive evaluation of Simple IP versus Mobile IP service and concluded that the Simple IP service must be rolled out first together with a marketing campaign that focuses on Mobile IP, which should be available shortly thereafter. The main reason for this decision was the absence of Mobile IP-capable handsets and the immaturity of Mobile IP clients offered for both Windows and Linux operating systems. It was also recognized that a Simple IP access network, as designed and supplied by a vendor of choice, was capable of supporting service offerings roughly similar to Mobile IP.

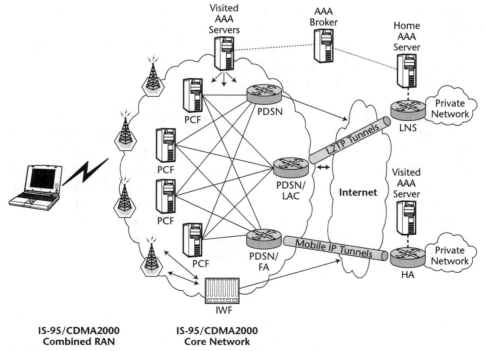

**Figure 7.10**   ACME USA combined IS-95/CDMA2000 core network.

The solution offered by the vendor to address Simple IP deficiencies included a fully meshed network of PCFs and PDSNs, which would allow Simple IP subscribers to remain connected to the same PDSN even if the serving PCFs were changing. This would ensure that the IP addresses assigned to the mobiles remain constant for the duration of the session, allowing for voluntary tunneling and other IP address-sensitive applications. Along with a packet data network, ACME Wireless deployed a distributed AAA network and elected to host an AAA broker service in one of its data centers. It also planned to start marketing AAA brokerage service to other carriers simultaneously with the rollout of Mobile IP service.

> **NOTE** ACME USA decided to support a combined public and private IP addressing scheme to facilitate versatility in offering of advanced IP services including upcoming Mobile VPN service.

Recognizing the complexity and immaturity of the Mobile VPN market, ACME USA decided to test-market a wide array of VPN options, including both voluntary and compulsory offerings. ACME USA offer for voluntary VPN users includes public routable IP address combined with service level guaranties, bandwidth management options, and different levels of security. Optional IPSec proxy service is also available and marketed to business customers such as partner ISPs as a part of remote wireless access outsourcing offer. The Customer Premise Equipment (CPE) program for voluntary VPN customers includes a variety of IPSec and PPTP software clients combined with support and location-based services.

Compulsory VPN service is also offered in an attractive package combining application hosting and proxying, address assignment outsourcing, and an extensive CPE (hardware/software bundles) program for enterprises including LNSs, IPSec gateways and HA equipment (for the soon-to-be-available Mobile IP service). An IPSec option is offered in combination with firewalling and bandwidth management and branded as a premium package, targeting large corporate accounts and government agencies. Both Simple and Mobile IP VPN services will be offered in standard form and rely on L2TP and Mobile IP tunneling, RADIUS AAA brokerage, and pilot SLAs with top-tier corporate customers.

# Summary

In this chapter we provided an analysis of the way CDMA2000 networks can be used to offer VPN services. Mobile IP and Simple IP CDMA2000 access methods based solutions have been analyzed, and address management and AAA aspects have been deeply analyzed. We also discussed related topics such as various HA allocation aspects. In the following chapter we will offer an overview of the equipment and platforms that can be adopted to implement MVPN solutions.

# CHAPTER 8

# Mobile VPN Equipment

In this short chapter we concentrate on the MVPN physical building blocks rather than on underlying technologies and systems architectures. We consider two major categories of MVPN products required to implement a workable system, MVPN clients and MVPN gateways, which can often be supported by or collocated with wireless data platforms and other devices. This chapter completes our discussion of MVPN and transitions to the last chapter of the book on future trends and scenarios in wireless data services.

A detailed review of the architecture of MVPN platforms and clients could easily span not one but a series of books, so our focus in this chapter is simply to analyze how different types of VPN components are used and their suitability to the task, rather than the details of their internal architecture.

## MVPN Clients

As opposed to gateways, which serve the needs of the whole or part of the network they represent, VPN clients are designed to satisfy the networking requirements of a single host. VPN clients might be therefore implemented on any type of computing device, from mobile phones to laptops to specialized equipment such as, barcode or credit card readers and smart appliances. VPN clients designed for a mobile environment must also satisfy requirements specific to mobile devices, such as optimized utilization of often-limited computing resources and power consumption and user interface adaptable for small screen sizes. These requirements are especially important when VPN clients are installed or incorporated into PDAs, smart phones, and other compact form factor devices.

**NOTE** The discussion about VPN clients in general is conducted in the context of voluntary MVPNs based on end-to-end tunnels or tunnel chains, since network-based VPN generaly does note require client functionality.

## MVPN Client Implementation

MVPN clients enable mobile devices to establish and support VPN communications channels. They have multiple functionalities, including tunneling, authentication and authorization, and security options. MVPN clients can be implemented in both software, which must work in conjunction with mobile device operating systems, and hardware, which must be either physically connected to a mobile device through an external or internal interface (such as a PCMCIA) or be permanently built into the device itself during its manufacturing.

### MVPN Client Functions

Voluntary Mobile VPN connections are most often initiated by the user, which often requires relatively few access control, authentication, and tunneling options. For this reason, the functionalities of a VPN client—especially in mobile environments—often constitute a subset of functionalities of the VPN gateway, discussed in the following section. It is customary for MVPN clients to support only one or two types of tunneling technologies, such as IPSec or PPTP and L2TPclients.

**NOTE** Often, wireline VPN clients are named according to the tunneling technology they support (e.g., IPSec client). That trend may be changing to a less technical convention in an effort by equipment manufacturers to make technology as transparent to the end user (many of whom are not technology-savy) as possible.

Client support for authentication is usually dictated by the VPN gateway and the rest of the infrastructure implemented in the private network the device is set up to access. Contemporary corporate networking environments, for example, tend to rely on RADIUS authentication combined with secure token cards (the so-called three-factors authentication methods versus two-factor authentication, based only on user name and password). A VPN client used in mobile environments may be mobility-agnostic where the underlying mobile network handles mobility transparently, or mobility-aware, as in the case of Mobile IP-based clients.

Within the Mobile IP clients, there may be a class of them supporting automatic fail-over to collocated care-of-address if a Foreign Agent was not found, and also automatic network interface selection (based on signal strength).

### Software-Based Clients

This type of VPN client implementation is by far the most widespread for both wireline and wireless VPN. Generally, software-based clients are cheaper to implement, easier to distribute and support, and can be removed or upgraded as needed. All of these advantages are especially appealing to mobile environments, but they come at a price. The software-based clients are usually less efficient and thus more processing-power hungry, which leads to potentially faster battery drain for low-powered mobile devices. For example, software-based 3DES encryption is known to be significantly less efficient than hardware-based implementations, reducing the performance and increasing user data latency. Software-based hashing functions such as MD-5 or SHA-1 are also capable of affecting performance and latency. These effects, combined with an unreliable air link, can have serious consequences on the user data aggregate bit rates.

Software-based clients are usually designed for specific operating systems. They can either be integrated into an operating system or implemented as a "shim" between the link layer and network layer of the OS protocol stack, also known as " bump-in-stack" implementation (see Figure 8.1). Some commercially available operating systems, such as Windows XP and Pocket PC 2002, have VPN clients built in. The OS-integrated VPN clients further simplify distribution and partially offload the burden of installation, distribution, and remote user support from IT administrators.

Software-based VPN clients can be implemented on both the OSI link layer and the network layer. A widely used example of a link layer VPN client is a PPTP/L2TP client supplied with the Windows 2000 operating system. However, by far the most widespread VPN clients are IPSec clients.

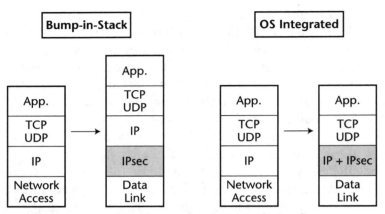

**Figure 8.1**   Bump-in-stack versus OS-integrated VPN clients.

VPN implementation in operating systems carries significant advantages, such as higher efficiency than shim implementation. On the other hand, the VPN clients implemented as a part of the OS are limited to features provided by the vendor and may lack capabilities required by different IT departments. In this case, you would need to resort to commercially available alternatives or in-house client software development.

### Hardware-Based Clients

Hardware-based MVPN clients are not commonly found today in the marketplace, although they may be expected to be implemented, in part spurred by the demand for more sophisticated mobile devices. Hardware-based clients can be implemented as add-on devices, such as PCMCIA cards, smart cards, or proprietary hardware connected to the mobile device through one of the available standard interfaces. They also can come on a chip or a chip set and be built into a mobile device itself during manufacturing. The latter option has a good potential to become the solution of choice for high-end mobile phones because of its frugal power use, potentially low-volume pricing, and other attractive qualities. Hardware-based MVPN solutions must be made firmware-upgradeable to keep up with changing standards and technology evolution. Note that an HW client normally would implement only a subset of the VPN client functionality (such as the most computationally intensive) and leave to SW application controlling the user interface.

# MVPN Client Design Issues

There are a few design issues unique to MVPN clients that we must consider. Most stem from the specific requirements of compact mobile devices and the nature of wireless communications.

### Limited Platform Resources

Engineers designing MVPN clients or adapting the existing solutions to various mobile devices, such as PDAs and smart phones, do not have the luxury of unlimited AC power and powerful microprocessors available for VPN client applications residing on stationary personal computers or UNIX workstations. One of the ways to address these constraints is to limit the options and functionalities supported by a mobile client. For example, tunneling support can be limited to IPSec tunnel mode ESP, and encryption options can be limited to 3DES and RC5. Also, the clients can be optimized for use with particular operating systems, such as Pocket PC or Palm.

### Unreliable Physical Environment

MVPN clients must be able to cope with an unreliable physical layer or wireless environment over which communication is taking place. When the wireless connection is unreliable (dropped circuit calls while traveling in a car is a good example), the end-to-end tunnels established between the client and the private network can often break and need to be reinstated. To address this problem, the VPN client must support options or procedures for fast VPN connection re-establishment, such as automatic login.

### Support and Distribution

Service providers, enterprises, institutions, and other privately networked entities deploying MVPN solutions must be mindful of the potential difficulties associated with the MVPN clients distribution, remote provisioning, and support. Mobile users tend to travel with their devices in different areas and might be hard to reach or limited in the ways they can communicate with centralized network support groups. Also, it is not feasible to dispatch technicians to constantly changing user remote locations for repairs in case remote support fails.

Distribution and installation of MVPN client software might present similar and even greater challenges to IT departments not only because of the constant mobility of the users but also the general immaturity of the mobile device universe. A remedy to address these problems might include a range of measures such as preventive on-site mobile device maintenance, outsourcing of mobile user's support to a third party, and broader involvement of equipment manufacturers in day-to-day support activities.

### Security Requirements

The security and authentication requirements for MVPN clients might often need to be more stringent than those of regular fixed VPN clients—and for a good reason. Stationary VPN clients are hosted on computers in closed quarters such as a house or remote office and are usually well protected from perpetrators. MVPN clients, on the other hand, are usually supported in mobile devices, which tend to be carried on a person. Such devices are often stolen, lost, and left unattended for extended periods of time. This makes them easy prey not only for property thieves but also for anybody wishing to get access to resources in a certain private network accessible by the device. Such issues of additional security can be addressed via a set of measures involving periodic user pooling, stricter account administration rules and mandatory use of SecureID cards, and other three-factor authentication methods.

## MVPN Gateways

MVPN gateways are the cornerstones of MVPNs. As opposed to MVPN clients, whose function is to satisfy the VPN connectivity needs of a single mobile host, MVPN gateways are the devices on which the MVPN infrastructure (often tightly integrated with the wireless data network) is built. Architecturally, in both wireline and wireless systems, VPN gateways, together with other networking elements such as firewalls, are located on the border between a public network (that is, the MVPN service provider network, or an ISP network supporting connectivity to the MVPN service provider network, such as in the case of CPE VPN gateway or CPE HA) and the MVPN customer networks. The MVPN gateway must also support some mobility scheme when it is collocated with a wireless data platform, thus creating complex multitiered networking environments.

## MVPN Gateway Implementation

The main task of an MVPN gateway is to ensure safe and secure transmission of the user data between mobile VPN clients and the destinations within stationary (virtual) private networks. This is usually achieved by selecting a robust hardware platform and implementing tunneling, access control, authentication, security, and other functions such as routing and various policy decision and enforcement mechanisms. Unlike MVPN clients, which usually support a limited set of these technologies, the MVPN gateway must be designed for versatility and hence support a variety of networking technologies and possibly different VPN client types (Figure 8.2).

MVPN gateways must be able to authenticate mobile users via multiple authentication mechanisms and interface to AAA servers via AAA protocol client such as a RADIUS client, terminate tunnels originated by MVPN clients (in voluntary VPN case), or by the wireless infrastructure nodes (such as Mobile IP capable PDSN), and establish secure communications channels by providing constant user data integrity and confidentiality protection. MVPN gateways must also support a variety of static and dynamic IP address assignment techniques, actually assuming the role of the network access server, and must perform additional tasks associated with tunnel termination and origination.

Like MVPN clients, MVPN gateways can be implemented in both hardware and software on a dedicated platform, or they can be combined with existing devices performing other tasks within a network, such as servers, routers, wireless data network element platforms, or combinations of these. As a result, often the nontrivial duties assigned to a typical MVPN gateway must be accompanied by the capability to handle routing and forwarding, buffer management, packet scheduling support, packet marking and metering, accounting and performance monitoring, lawful interception, various data mobility schemes, and other typical wireless data networking functionalities. Such a variety of functionality has positioned network devices known as IP services switches (more on this in the next section) as the platform of choice in the carrier environment to support MVPNs, in part because of the service richness they can provide. Typically they not only support VPN termination, but also subscriber management, user services customization, firewalls, NAT, customer network access link termination, customer self-provisioning, and monitoring capabilities.

**NOTE** The downside of this feature richness is the potentially high cost of such devices, some of which have already been implemented by a number of vendors. However, the cost of the device may be often balanced by its flexibility and revenue generation potential.

Both dedicated and collocated MVPN gateway implementations can be *native*, where VPN gateway functionality is integrated with the platform operating system and often the hardware, or *bump-in-wire*, analogous to bump-in-stack implementation of MVPN clients, where a VPN gateway implemented on a dedicated platform is collocated with a router and attached to it via a physical interface. The specific case of devices combining MVPN gateway and wireless data platform functionalities deserves a bit more detailed evaluation, since we expect that this type of equipment has a good to chance to become a dominant way to implement MVPN gateways.

## MVPN Gateways and Wireless Data Platforms

Before examining the MVPN gateway/wireless data platform combinations, we need to analyze the main types of wireless data platforms, their characteristics, and their suitability to VPN gateway functionality support. This analysis will be brief, however, mainly because the detailed discussion of modern router, IP switch, or an emerging class of wireless packet data platforms, for that matter, might easily evolve into not one but a series of books.

Wireless data platforms can be implemented in a variety of ways. The wireless data platform implementations can be roughly classified in two major types:

- Router or IP switch-based
- Server or general-purpose computer-based

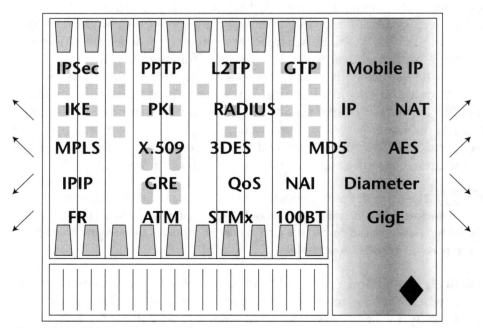

**Figure 8.2**  MVPN gateway functions.

Each has different properties as well as unique advantages and constraints. Servers are *general-purpose* computers optimized for serving the needs of a network or a group of users rather than a single user. Routers are in fact *special-purpose* computers optimized to forward data packets from one physical interface to another.

We must admit however, that the distinction between these two approaches is very fuzzy and has become even more blurred in the last few years. One way to clarify the distinction is to look at the platform operating system rather than its hardware. Traditionally, routers and IP switches were implemented based on specially designed real-time operating systems (RTOSs), such as VxWorks produced by Wind River systems, while servers relied on general-purpose non-RTOSs, such as UNIX, Linux, Windows NT, and Solaris OS. This approach, however, recently also became less applicable because of newly introduced dedicated routing platforms for wireless data and other types of networking relying on non-RTOSs, which have been simplified and made to conform to RTOS requirements (such as Starent ST16 wireless data platform, based on the generic Linux operating systems) and general-purpose computing systems based on PSOS and other RTOSs.[1] Other examples include vendors introducing complex routing and tunneling devices based on clusters of generic, often off-the-shelf computing systems (such as the now-defunct Cambia Networks, which developed PDSN, GGSN, and HA based on off-the-shelf Linux server produced by IBM), as well as dedicated router designs equipped to support functions such as firewalling and application layer processing, traditionally performed by servers.

Despite these difficulties in classifying wireless data platforms, most of them can still be defined—if increasingly loosely—and segmented into two classes, allowing us to analyze the pros and cons of each:

- *Server-based wireless data platforms* are based on generic computing hardware and operating systems originally designed to support a variety of general *computing* tasks and applications, software-enhanced to support networking tasks such as packet forwarding or routing, tunneling, and AAA functions.

- *Router/IP switch-based wireless data platforms* are based on hardware and software dedicated to performing a variety of *networking* functions (residing on the lower four OSI layers: physical through transport). Architecture of such platforms is usually optimized for packet data handling and forwarding or switching and tunneling and is not normally intended to support any applications outside of data networking.

---

[1] An interesting example of the latter is an infamous first-generation unmanned Mars Pathfinder rover launched by NASA in 1997 (for further information on this unique vehicle, see http://mars.jpl.nasa.gov/default.html and http://nssdc.gsfc.nasa.gov/planetary/mesur.html).

Both of these platform types have certain advantages and constraints when used in wireless core networks to support a variety of tasks including MVPN. Let's consider these two types, along with several subtypes, in more detail.

### General-Purpose Computing Platforms

Computer systems designed to perform a variety of tasks rarely excel in a particular one, forwarding or encapsulation of data packets being no exception. Most computer systems are designed to meet general market requirements and maximize their computing performance, while staying under budget both financially and technically in terms of system resource allocation.

For example, computing platforms optimized to run OSs like UNIX and Linux have been designed for data processing and general computing tasks, with resource allocation strategy not optimized for such tasks as routing, PPP termination, and especially encryption, compression, packet fragmentation and reassembly, and tunneling and tunnel switching. The memory hierarchy of a typical computer designed to run UNIX OS or its siblings, such as SUN SPARCstations, relies heavily on caches or virtual memory residing in nonvolatile memory, such as hard drives and other types of relatively slow-access mechanical devices. While cost-effective and suitable for general computing tasks, this approach often is not the most reliable, since mechanical devices are more prone to failure, nor is it optimal for running mission-critical applications such as routing that require frequent fast local memory pages.

Then there is Input/Output (I/O). In general computer systems, the I/O is a necessary evil often looked down upon by designers. In contrast to routers and other networking devices designed to deliver packets from one interface to another, I/O interfaces are what all the hard work is being done for.

> **NOTE** These and many other problems with UNIX-like operating systems in part stem from the fact that their network layers were originally designed for low-speed protocols such as SLIP, which used to be mainstream decades ago, and not for today's high-speed technologies and large-scale packet encapsulation and forwarding.

When introduced to the task of handling complex wireless packet data functions entailing routing, switching, and encapsulation, such devices' packet forwarding and throughput constraints would not allow for efficient functioning in high-performance networks often spanning continents and hundreds of thousands of subscribers. Moreover, rather limited interface support options as well as inflexibility of hardware design, which often disallows hardware-based packet processing and other functions such as compression and encapsulation typically required by wireless packet data systems such as GPRS and CDMA2000, make these solutions only suitable for proof of concept, trials, and initial rollout stages of commercial deployments. In such small-scale deployments, however, these platforms often make sense, offering cost-effective alternatives to bringing purpose-built and therefore expensive routing devices in unproven and therefore high-risk markets.

### Routers and IP Switches

The first breed of computers optimized for packet forwarding, named routers, was introduced by Cisco Systems in the late 1970s. The router was conceived to address the problems associated with generic computers when applied to networking tasks by creating specialized hardware often combined with real-time OSs optimized for data packet forwarding and other networking functions. Introduction of routers was a significant step forward for telecommunications and marked the beginning of a multibillion-dollar industry now known as data networking. Today routers come in many shapes and sizes, with different specializations and even under different names, such as an "IP switch" or a "network appliance." Not surprisingly, the majority of the successful wireless data platforms, while often based on specialized architectures, are still built around the familiar routing engines.

Originally, routers were designed to dynamically communicate with large numbers of networks via a limited number of links. As data networking technology progressed, to accommodate for the support for new functions such as subscriber access link termination, deep packet lookup, and, later, VPN functions and wireless data services, modifications to the original routing concept were needed. For instance, *remote access servers* (RASs) were introduced to handle subscriber termination, and *IP switches* were introduced to handle VPNs and other advanced data services such as labeling at the edge of MPLS networks, NAT, firewalling, and IPSec and tunnel switching. The latest wireless data platforms tend to be based on traditional routers, RAS devices, IP switches—all of which have their own advantages and limitations regarding mobility business and technologies like MVPN. Later in this section we will devote some attention to each of these.

To qualify for support of wireless packet data services in general, in addition to its traditional ability to efficiently forward packets, a routing platform must satisfy certain requirements:

- The ability to terminate subscriber traffic
- The ability to encapsulate subscriber traffic on a large scale with minimum delay

- A full set of carrier class features suitable for the Central Office environment
- Support for significant volatile local memory, such as dynamic random access memory (DRAM) resources and fast memory paging
- Support for wireless system-specific signaling and logical interfaces
- Support for QoS technologies such as DiffServ
- Support for packet data mobility schemes such as GPRS and Mobile IP

In addition to these requirements, a wireless data platform aspiring to become an MVPN gateway must also support the following:

- Dynamic tunnel switching (see Chapter 5)
- Multiple tunneling protocols (see Chapter 2)
- Dynamic policy provisioning (COPS, LDAP; see Chapter 5)
- Encryption, compression, and AAA client

In the carrier environment, the wireless data platform must also support the following:

- A variety of I/O modules to interconnect the device to the mobile network and to customer networks.
- Ability to interface with a large number of customers' networks supporting overlapping private IP address space, as well as multiple virtual routing tables or full-blown virtual routers customers may manage.
- High availability and other carrier-class features.

### Traditional Routers

No doubt, traditional routers are more appropriate in this environment than generic computer systems, since efficient routing is, of course, their strong point. Carrier class features can be easily added by designing a robust chassis with various redundancy mechanisms. Tunneling and labeling capabilities can also be achieved in both hardware (for example, by designing new "encryption accelerator" cards) and software (by optimizing protocol stacks). Tunnel switching and scalable subscriber termination might, however, create certain problems and prompt significant architectural changes, which might potentially cause the router to evolve into something else, like an RAS or an IP switch. Mainly for these reasons, classic routers, when used as wireless data platforms and MVPN gateways, often suffer from low scalability and a lack of essential features. IP service switches or advanced edge routers utilizing virtual routers (see Chapter 5), network processors, and other advanced techniques are likely to be more suitable for the task of supporting MVPNs in carrier environments.

### Remote Access Servers

Remote access servers were originally introduced by companies like Livingston, Ascend Communications (both acquired by Lucent), and 3Com (now CommWorks after the spin-off) to address the need to terminate large quantities of dial-up subscribers using such link layer protocols as PPP and Serial Line Internet Protocol (SLIP). RAS is a combination of a routing engine and modem banks (often implemented as a farm of Digital Signal Processor, or DSP, chips), software-optimized for subscriber termination, and additional memory necessary to store PPP state information. Architecturally, RAS is used to terminate subscribers accessing a private network or the Internet and aggregate high numbers of low-speed dial-up communications sessions or ISDN calls established over a PSTN into time slots on a small number of high-speed interfaces such as T1s or T3s. While quite appropriate for wireless circuit data systems based on dial-up connections (CommWorks' Total Control chassis is a popular example), RAS might be less well suited for wireless packet data systems and MVPN support mostly because of its already high complexity and cost and hard limit on the number of subscriber sessions (usually a few thousand) and session bandwidth.

### IP Switches

IP switches, a new breed of routing device, are based on a routing engine or a cluster of routing engines and specifically designed for scalable packet subscriber termination, via DSL or cable modem, and encapsulation, as well as other tasks beyond the capabilities of traditional routers or RASs. The recent examples of successful IP switches include Shasta 5000 from Nortel Networks and Springtide 7000 wireless from Lucent Technologies. The physical interfaces of the IP switches can be shared by a high number of subscriber sessions, which can be scaled up and down with the interface bandwidth to the desired amount of memory and processing power for the routing engines and network processors. Tunneling and processing the intensive tunnel-switching capabilities of IP switches are usually supported in hardware for higher efficiency. The combination of these features, originally developed to address the requirements for scalable VPN support, is well applicable to the support of both wireless packet data and MVPN functions, making IP switches the best candidate MVPN platform.

## Summary

In this chapter we provided a high-level discussion of the architecture and design of platforms that support MVPNs and routing devices which can be tasked with the support of wireless data. End-to-end support requires the MS to support VPN client functionality—which may or may not be specialized to handle mobility—while the network must support termination of mobility protocols related to VPN tunnels, AAA functions, and possibly user mobility management if the clients are mobility-aware (as in the case of Mobile IP-based VPN clients). In the next chapter we conclude the book with a look at where the wireless data industry may be heading.

# The Future of Mobile Services

The future of mobile systems and services is not easy to predict. Many factors may influence its evolution, and we don't know and would not attempt to *predict* how current trends in the support of packet data communications and applications will play out with the expected integration of a pervasive mobility environment and a variety of computer systems. In this chapter we instead *evaluate* these trends and their directions.

An example of evaluating the current situation to see what the future may hold can be found in the definition in 3GPP of application programming interfaces (APIs) created for quicker development of mobile applications. Aside from the safe observation that standard interfaces for applications developers such as Parlay API or Web service interfaces will become widely used, it cannot be predicted with any certainty how the services themselves will evolve in wireless systems. Most likely each provider will offer a set of nonportable applications that can't be recreated in another carrier's environment. This in principle will reduce customer churn and attract new customers interested in the particular set of applications. This approach may seem obvious, but it is a radical departure from the current wireless systems paradigm where almost all providers offer the same or similar services.

In the future, mobile data networks are most likely to be *content-* rather than *connectivity*-driven. Today's mobile networks offer connectivity service to a person or to a machine, with the content transferred almost always transparent to the wireless service provider. However, in recent times, we have witnessed the trend to offer value-added services to customers above and beyond connectivity, like traffic conditions reports based on location or explicit user query, and stock-market-related services and alerts. These services have been also offered using Wireless Application Protocol (WAP), as well as the classic voice-based or Short Message Service (SMS) interfaces. Often the user experience in these services has been less than stellar, so the expected high take rates were not achieved (perhaps with the exception of SMS), most likely because, fundamentally, customers want a more user-friendly approach to consultation of information. It is expected that with the introduction of packet-switched data and the associated upgrade of user equipment computing and display capabilities and location- and user-profile-based services, users will be presented with a friendlier and more attractive set of applications and easier to use and more functional mobile devices.

Another aspect we consider in this chapter is which services are more likely to be successful. We identify three main categories of services: *person-to-person, person-to-machine,* and *machine-to-machine.*[1]

Packet data capabilities pave the way for a new class of applications, such as those based on communication between hosts of data applications. In fact, in the future it may be possible for many smart appliances (fleet management devices located in delivery trucks, vending machines, water meters, and so on) to have a low-cost wireless data interface through which they can exchange information with remote computers located in fleet management centers, help desk and diagnostic centers, credit validation and authorization centers, and information repositories or software repositories for upgrade downloads. Mobile phones themselves could be used for automated toll and parking fee collection, and could receive upgrades and personalization data via their wireless interfaces. Among others, broadcast and multicast service support is being standardized in 3GPP, and they will surely foster the birth of new applications based on the delivery of the same content to multiple users at the same time—content that in the past could not be delivered over cellular wireless networks in a profitable manner.

In addition to service categories discussion in this chapter we present possible scenarios based on past developments, current trends, and predicted market directions. We begin with current standards directions in the industry, then we evaluate possible development scenarios (MVNO-specific included) and the roles advanced services such as Mobile VPN would play in them. We conclude the chapter with an analysis of WLAN deployment

---

[1]Note that the latter is a class of services that is not very common in today's wireless networks.

trends and their business and technical positioning in the industry versus current cellular systems (competition versus convergence).

## Current Wireless Systems Industry and Evolution of 3G Systems

Historically, many of the world regions have been dominated by protectionism—the desire to impose barriers to free movement of people, information, goods, and capitals across regional boundaries, for the purpose of protection and sustaining local economies and thus preserving the wealth of countries. Among other negative consequences this has also resulted in the creation of incompatible wireless systems and spectrum allocation strategies in different regions of the planet. However, recently two factors are changing this traditional model: product development and trade are happening more and more on a global scale and a global Internet makes the exchange of information simple and often free.

As an increasing number of people enjoy the ability to freely cross borders for business or leisure while requiring basic services such as email, financial trading and banking, news, and virtual messaging, we need to find a way to integrate or at least harmonize wireless systems—if only from the user's experience point of view, that is, from the services and applications perspective. In part responding to this situation, in April 2002, 3GPP and 3GPP2 partners met in Canada and reached an agreement to follow a common path in the development of the 3G IP-based systems' evolution. The ultimate goal was to allow global roaming across technologies (and thus across cellular wireless standards). Thus the ability to support services access from multiple technologies has been recognized as an important requirement for 3G systems. Transport convergence is one of the main reasons behind the success of IP in wireline networks, which allows internetworking across different networks. It is interesting to observe that the convergence of wireless systems is happening as well, but this time it is happening on a *service level*.

How will this transition occur? Initially, the access experience will be discontinuous. Users will be signing on and off services many times a day, each time potentially changing the type of access network available and reestablishing connectivity. Later, with wider adoption of the IP layer mobility technologies such as Mobile IP, the mobile user will be permanently able to access services without the need to reconnect in a multi-access environment. The driver for this will be in the applications and services, which will require permanent IP layer reachability to be supported (keeping IP address constant, for example). As a result, converged wireless data systems will likely go through an early-adolescence period and will not support uninter-rupted connectivity (mentioned in Chapter 1). Until then, global IP level roaming (provided via RADIUS or DIAMETER protocols or based on digi-tal certificates) is expected to cover most mobility needs.

## Service Aspects

In Chapter 1, we discussed how 3GPP- and 3GPP2-defined systems are evolving toward a common IP-based core network to support multimedia services. We use the term "multimedia," since it is the original language used by the standards bodies. However, multimedia support is not the only goal of IP-based services capabilities. Rather, the goal is the integra-tion of data and other media onto a single uniform *session-handling* system. This approach will allow for novel service experiences and the integration of *media-based* interaction with humans and machines with *data-based* inter-action with humans and machines, resulting in an innovative ways to present services to a user equipped with mobile devices radically different from today's mobile phones

**NOTE** This new paradigm, however, will not exclude the existence of simple and single-purpose devices like IP phones, telemetry appliances, pay toll, or vending devices.

The architecture of all-IP-based systems, or IP Multimedia Subsystems (IMSs), shown in Figure 9.1, is based on the Home Subscriber Server (HSS). An enhancement of the Home Location Register, HSSs support subscriber data related to IMSs based on Session Initiation Protocol (SIP—[RFC2543]) servers that control media sessions establishment and gateways used to interwork with other systems such as the PSTN and legacy mobile systems. The HSS will include functionality previously associated with the HLR, including wireless network access control and subscriber profile manage-ment, as well as multimedia services subscription data and IP-based regis-tration information. The registration information sets up a mapping between user public identities and the services that the user has currently activated (for instance, presence or multimedia conferencing services nicknames), the user private identity (such as the IMSI), and the current IP address used to reach the user. User registration to services depends on explicit authentica-tion, subscription, and possibly credit checking. The interface to the HSS in 3GPP is based on MAP (for wireless access network access control) and DIAMETER, while in 3GPP2 it is only DIAMETER based.

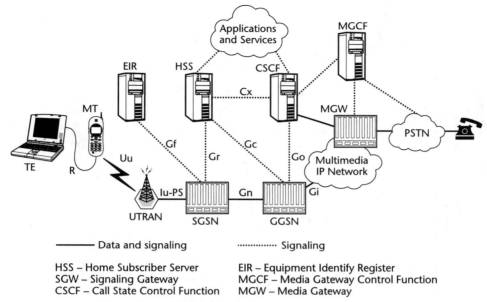

**Figure 9.1**   Simplified IMS architecture.

Figure 9.2 outlines the registration process for a subscriber accessing IMS services in 3GPP-compliant networks (Release 5). The subscriber equipped with a proper terminal first attaches to a UMTS network. Then a Packet Data Protocol (PDP) context is set up for an IMS-enabled APN (associated to an IPv6 network hosting IMS servers). Then it uses the services of a Proxy CSCF (Call Session Control Function, a SIP proxy server defined for 3GPP IMS architecture) to register with the HSS in the home network. In the process, the home network assigns the subscriber to a Serving CSCF (a SIP server defined in the 3GPP IMS architecture to offer service control and session control to subscribers). The S-CSCF is located in the home network, and it can therefore deliver the same look and feel of services independent of the current user location and serving network.

The IMS uses packet-based services. At this time, packet-based services do not include efficient support of multicast, which is currently supported via replication of the multicast information over unicast bearers at the gateway nodes (the GGSN). A set of specifications is being readied for 3GPP Release 6, so that the radio interface will become multicast and broadcast services capable, and a new set of services will be made available.

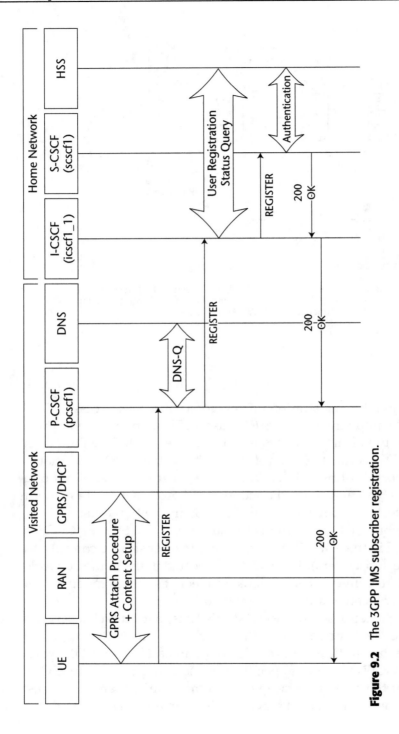

**Figure 9.2**   The 3GPP IMS subscriber registration.

This creates significant new revenue potential, since the availability of IP-based broadcast services will certainly make it possible to generate revenues from advertising in ways unknown to current networks. This will create a demand for multicast and broadcast content distribution networks, which will need to track user mobility. This in turn will drive a new kind of MVPN, namely multicast MVPNs, where traffic is sent only once to any IP network access point and the access point has at least one member of the multicast group. This can also be accomplished simply by extending current multicast routing protocols, or by using multicast tunnels between the multicast distribution gateway and the access points.

IMS will be IPv6-based, and it will potentially aid the large-scale introduction of the IPv6 protocol in carriers' networks. As a result, the IP VPN technology, which needed to support IMS VPNs—such as those associated with MVNOs or large organizations—will need to be IPv6-aware. In our view the use of MVPN will still persist in IPv6 networks, despite the disappearance of its need in real private address space. In fact, tight access control is a prerequisite for commercial data service delivery, which makes it necessary to use MVPN technologies to govern access to services at the network layer.

Finally, the ability to deliver media flows comes in 3GPP, recently 3GPP2 started some activity on this too, with the ability to bind a UMTS network bearer to the related media session via a COPS-PR-based [RFC3084] policy control mechanism that uses the delivery of media authorization tokens to the session endpoints. These tokens are then used during the bearer allocation request to validate the association of the bearer to the media stream and to install the associated packet filters, thus allowing service providers to charge simply based on media session duration or content and not for the bearer service. This feature is supported by the so called Go interface between the GGSN and the Packet control function of the CSCF

## IP-Based Mobility

IP mobility support is an increasingly popular method for keeping a mobile station connected to the Internet while changing the point of attachment to the Internet. However, sometimes a mobile station isn't required to be permanently reachable at the IP layer, so this kind of mobility support may not be part of the evolution of mobile systems.

The IETF Seamoby (for Seamless Mobility) Working Group has drafted a different proposal addressing this issue. The MS would not be involved in explicitly registering with the HA; rather, the network tracks it so that the

MS is always reachable at a tracking server, which then could page the MS when some entity need to communicate with it. We believe this may not prove to be a successful approach, since most access networks come with this capability at the link layer. More interestingly, when multiple IP level handoff candidates exit at a given location, the problem exists on how to detect the *best* handoff candidate. This problem is somewhat philosophical, but it has both business and technical implications, such as roaming to another provider network, choosing a low-cost option, and using an access point that allows for service continuity with unchanged QoS. The Seamoby WG is also considering a proposal on how to address this problem. This proposal from the Seamoby Working Group faces an enormous set of unresolved problems that would limit its practical deployment, if the standards were approved.

There is an urgent need to make some methods available for mobile user authentication in an IP-based network by just using IP to exchange authentication information, with the access point acting as a network access server. In the IETF, the Protocol for Carrying Authentication for Network Access (PANA) Working Group has been formed to address this issue. Many proposals are on the table—some that reuse DHCP authentication [RFC3118] and others that extend EAP [RFC2284] to be IP based, in accordance with some recent trends in 802.11. The PANA WG promises to develop a universal authentication protocol that may be used in any network to allow for universal and seamless roaming. However, it may well be that the industry—especially driven by proliferation of WLAN applications—will converge on another de facto standard if the PANA WG is not quick enough to find a solution.

## Billing for Wireless Data Services

Originally, billing for wireless data services in the cellular environment was based on the innovative concept of volume-based billing. This model was expected to be more attractive to the consumer than the traditional time-based usage metering, which kept a connection on continually and used it only when necessary to transmit and receive data. Service providers saw the added benefit of encouraging limited use of radio resources while offering high-margin data delivery services. However, this model has proven unsuccessful for the general consumer market, other than for transactional applications such as those offered via i-mode in Japan.

**NOTE** i-mode is an NTT DoCoMo proprietary technology, currently licensed in Europe to providers such as TIM and KPN, that is based on the delivery of services via a gateway node to partners offering wireless data applications ranging from gaming and dating to mobile banking. This service has attracted millions of subscribers in Japan, and it is based on a small fee for the transmission of units of data volumes. The wireless operator also certifies partners entering in agreement with them (the *official i-mode sites*) and acts as a micropayment collection center on behalf of the official i-mode sites, in addition to other transactional and presentation services (such as the inclusion in the i-mode menu appearing on the i-mode terminals, allowindg users to avoid the need to type in the site URL). Other sites may also be implemented Web pages following the i-mode compliant subset of HTML (Compact-HTML), so that i-mode users can access them. These sites not entering agreement with wireless operators are called *unofficial i-mode sites*, and the user is required to type in their URL using the mobile device man-machine interface in order to reach them.

Nevertheless, beyond its use in i-mode, volume-based charging in general is not considered to be an attractive option for consumers and corporations alike. In fact, unless volume fees become sufficiently low, there is no acceptance of paying hefty amounts of money to transfer documents or multimedia over the wireless link. On many occasions, the cost of this may surpass the productivity gain or time value of having ubiquitous access to information, making this business model not viable, since users would rather wait to access information until they are at some location allowing them to use an alternate network access technology that suits their needs and finances better.

As a result, one expected trend is for wireless carriers using the flat-fee approach already adopted by the WLAN access providers. This will attract the masses to the wireless network data services experience and let them become familiar with and pay for an ever-growing host of applications, offering unique value needed to compensate for the commoditization of voice-based services. Competition will not be solely in the access fees, but also in the ability to deliver content and services that are valuable to customers. Sophisticated application level billing will be an integral service, as well as the collection of billing information from partner ASPs, and the correlation of network access level Charging Detail Records (CDRs) with application level CDRs.

Finally, credit management via prepaid or credit card accounts may be extended with the integration of financial or banking services offered by the wireless carrier, who may tap into the financial services market, perhaps with the aid of some existing online merchant services or specialized online banking partners, and offer micro-payment service for a multitude of everyday purposes.

## The Future of Wireless Service and Systems

Let's consider possible service scenarios, depicted in Figure 9.3, as well as the players and technologies that would enable them. A critical requirement of Internet-based services is the ability to deliver an environment the user feels comfortable with and is attracted to. This environment may be created via human-machine interaction: a well-defined graphical user interface (GUI), easy-to-use service options, interface customization, and self-adaptation to customer usage profiles. It may also involve hosting communities of interest, where groups of individuals can share emotions, experiences, ideas, and information. This requirement must be satisfied by person-to-person and person-to-machine data services offered by wireless systems, but of course, not by machine-to-machine applications. In this section we examine these aspects of future wireless systems, as well as new services offered by some new types of carriers entering the market, such as MVNOs and WLAN access service providers.

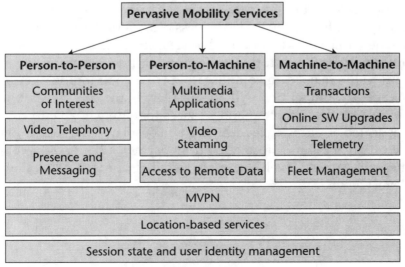

**Figure 9.3**   Future mobile services examples.

## Person-to-Person Services

In traditional telephony, interactions between two or more human users of the service usually occurred after some mutual knowledge had been established and after the exchange of phone numbers via channels other than the telephone network itself. We refer to this as an "out-of-band method." This model is still valid today, with the exception of consumer-to-business (e.g., free phone calls, dialing of a number looked up in yellow pages directories) and business-to-consumer (e.g., telemarketing applications and customers surveys, which are not frequent in mobile telephony, other than from service providers themselves) applications where only one party knows the other in advance. Dating services removed the need to exchange out-of-band phone numbers, although some of them rely on SMS-based exchange of information.

In next-generation person-to-person services, phone numbers will cease to be the only identity through which users will be known. More likely, users will subscribe to communities using a public identity. They will establish social contacts with members of these communities, creating a comfortable environment and fostering demand for other mobile telecommunication services. For instance, a user may create a virtual community involving all of his family members for the purpose of exchanging text messages, images, short videos, and voice conversations. Corporate communities of interest will also be set up to offer messaging and information dissemination services. Information distributed within these corporate communities may be privileged or restricted. Community membership rights will be critical in this type of environment. Managed community services may become an important part of a wireless carrier business, frequently bundled with managed VPN for security and privacy.

Community members may be alerted via an icon-based interface when the others are reachable. In this scenario, phone calls will be placed only when another user is available; otherwise, other media-based messages can be left in users' mailboxes. Users will be able to selectively signal their availability to other members or member subsets depending on, for example, the time of day and business status. In this approach, a set of addresses would no longer be stored locally in a phone book; rather, a list of the "friends" or identities organized on a per-community basis would be stored in the network. Of course, this does not mean the death of the phone number, since the old-style phone book is likely to exist in order to place and receive phone calls. Instead, this model offers new ways to let individuals interact through mobile and nonmobile devices. This will significantly increase the amount of information exchanged via other media [such as picture messaging, already being rolled out in Europe via Multimedia Messaging Service (MMS)], and consequently the service provider's revenues.

Another important aspect of this new user-interaction model is the entity acting as the communities' host. Users may in fact be able to subscribe to communities hosted by parties other than the wireless service provider. However, the wireless service provider may potentially offer much more added value that would encourage customers to use communities hosted in its environment, since the wireless carrier has the opportunity to integrate in its environment other valuable services such as location-based information, push-to-talk voice services, session related information and single sign on. Thus, communities hosting by wireless providers can result very competitive and highly differentiated with respect to what the third parties may offer. Third parties willing to offer similar user experience might eventually be forced to partner with the wireless carriers, becoming one of the partner ASPs, for example.

Another service that promises to become a mainstream revenue generator in 3G systems is video telephony. However, while it would significantly enhance interpersonal communication, pricing issues may prohibit its success. Historically, video telephony has not been successful in the wireline, which seems to be the case even now, when many households and businesses can be easily equipped for inexpensive high-speed communications. Perhaps mobility and the need to associate a voice with a face among the individuals involved in a conversation may prove to foster this service, along with the ability to distribute an image of a current location at the time the video-enhanced speech session is placed.

Initially, at least, community access should not be bound to any charge other than a modest flat subscription fee, in order to foster the growth of the services associated to them. As with many other telecommunication services, communities suffer from network *externality*, where the value of the service itself is bound to the existence of other subscribers using the same service. Access barriers should be minimized in order to attract users to communities where they can use a plethora of services, which together will constitute a significant revenue stream. As mentioned, the creation of communities and the ability to manage "friends" lists and customized user profiles within them is imperative for customer retention and attraction, since the social aspects of such unique services will have a significant effect on the users. Wireless carriers will therefore most likely be encouraged to create mutually incompatible communities hosting environments, similar to the instant messenger services offered by players such as MSN, Yahoo, and AOL.

## Person-to-Machine Services

Unlike person-to-person services, a person-to-machine service offered by wireless carriers does not entail synchronous, real-time interaction between humans. For this set of services, humans interact with hosts of data or multimedia applications, encompassing entertainment, information retrieval, and transactional applications. These services are based on the availability of "content" that humans are willing to access.

Users belonging to virtual communities can themselves generate some of the content, via person-to-person exchange of images, short movies, text, and Web content. Still, it is expected that most of the content will be provided by partners and hosted either on partners' networks or on wireless carrier facilities. This will require the use of MVPNs to enforce service exclusivity and protect against denial-of-service attacks and unauthorized access.

In this book, the term *content* has a broad definition; it applies to everything a user may experience by interacting with a person or a machine, when the machine does not act as a mediator between two human beings interacting in real time. For instance, a chat session, which still is mediated by an application server, is not content, but a Web page created by a user of a community or that is presented to a user during a travel reservation session or a weather forecast page consultation is. Success of content-based services in person-to-machine scenarios, as in the person-to-person case, depends highly on the user-friendliness and the clarity of language used to interact with the user. Very often, to the chagrin of the service provider, this detail is forgotten when these services are designed.

Person-to-machine services are used in today's networks in information products such as traffic conditions reports and interactive voice-response systems used to retrieve account status information or to update information, such as credit card number, in subscribers' profiles. Usage of these services is not yet widespread, since the majority of users still prefer direct operator assistance to the sometimes difficult-to-understand menu options or voice-response systems that fail to correctly guide the user. Clearly, machines are better accessed through point-and-click-based interfaces rather than via voice-based interaction, although there are indeed some applications based on Voice XML, a markup language similar to those used to create Web pages that is optimized for voice-based content navigation and collects voice input and executes voice commands. When mobile devices become equipped with enhanced graphics and point-and-select input, better human-machine interaction will be possible, leading to "enhanced user experience" based interfaces. One of the issues that must be considered is how the user inputs the information while interacting with machines. With the small form factor mobile devices, it is unlikely the user could easily input text, which requires a minimum amount of necessary text entry on a user-friendly interface.

From our view, person-to-machine applications are poised for success with both wireless subscribers and service providers for a number of reasons. First, these services are not expected to suffer from network externality, so they may be the first to be requested while community-based and other novel kinds of person-to-person services are still ramping up. Second, timely access to up-to-date financial information, weather forecasts, airport schedule and flight status, traffic reports, or location-based entertainment typically matches the anytime-available capabilities of information access media such as mobile devices, and the likelihood that the user is on the move while requiring this information is pretty high. Additionally, most of these services deliver value at well-defined time intervals, when the desired information is delivered or a transaction is complete, and it is therefore possible to define a fair price per application level events and avoid charging on a per-byte basis. Customers may be required to directly pay the wireless carrier, or the endpoints of the transaction may be required to pay a commission to the wireless carrier, or both. For instance, a bank offering mobile banking services may charge customers a fee per transaction while on the move. Or the bank may offer the services for free as a way to attract customers and then have a framework agreement with a wireless provider to deliver the service for a flat fee or a per-event fee.

Communities may become a vehicle for human-machine services. A community may be entitled to receive some information from the machine reserved just for its members; for example, a community may be defined as a group of customers of a big banking institution and any member may be addressed with news from the bank or welcome to interact with virtual members, which are simply machines offering help desk service or acting as virtual attendants.

Familiar Internet access service is another good example of human-machine application. In the near future it is most likely to be bundled with the wireless data service offer and not likely to generate significant revenues based on per-traffic-volume fees. On the other hand, wireless access may foster a breed of new services on the Internet itself, and these services may in turn use information derived by accessing data available from interfaces wireless providers offer to third parties, such as via OSA gateways or Web interfaces.

**NOTE** Open Services Architecture (OSA) is standardized by 3GPP in the TSG CN WG5 group, which defines a set of APIs that can be used to invoke services wireless carriers allow third parties to access at OSA gateways. Such services may include, for instance, session status information, user availability and location, and billing and accounting. The delivery of such information to third parties will constitute a revenue stream.

Another application that is thought by many as a potential driver of 3G services is video streaming—not unlike video telephony. In general, we expect video streaming to be offered in the form of short-duration clips that satisfy users' curiosity about a news item or events, or about the user's current location, such as historical news, cultural information, and tourist information. The delivery of lengthy entertainment video or entertainment clips that may be easily available at home via broadband access or cable service cannot realistically be expected in the foreseeable future mainly for the bandwidth limitations and other arguably obvious reasons such as device screen size and short battery life.

Let's now consider a hypothetical real-life person-to-machine service scenario. John and Alice arrive at their travel destination one evening without a hotel reservation. They book it swiftly via a location-based hotel directory service, check in, and decide they are ready for a night out at the cinema. They have already previewed a movie in the car before arriving at the hotel. They invoke a location-based service in their mobile device to decide what local cinema to pick among those offering the movie they are interested in. They check availability, select the seats, and buy the tickets on the way to the cinema. To enter the cinema, they just need to point their mobile device to a Bluetooth-enabled checkpoint that exchanges authorization information with their device, equipped with the subscriber's digital certificate that was used to provide the user identity when the ticket was sold. After watching the movie, they are reminded about a prearranged get-together with friends by their mobile calendaring application, so they send them a message showing their rendezvous point on the map, obtained from a simple request to a location-based service. They eventually go out for a late-night dinner at a restaurant suggested by their friends (not via an online service this time).

Even this simple, nonbusiness related example illustrates how many services can be provided to a customer and how many revenue opportunities can be made available to an operator if at least some of the new technologies we described see mass deployment. One thing that carriers must avoid is creating the expectation that all of these services are free, which has worked fairly well with SMS (that is, SMS customer expectation about the service fee were set before service introduction and accepted by the consumers), and as such it may be the same case for other wireless data services.

## Machine-to-Machine Services

Machine-to-machine is to us the most interesting and unexplored type of mobile service. So far, most of the services offered by mobile operators almost always involved a human user at one of the endpoints of the communication sessions. With the mass deployment of wireless data, this paradigm can now be shifted, and revenue streams from machine-to-machine interaction can start contributing to operator's bottom line.

The space for this kind of application is very wide and largely untapped. Examples include letting mobile manufacturers remotely upgrade handset software via the wireless interface, using the mobile as an authentication and authorization token for payment purposes instead of the traditional credit card, or bundling wireless terminal capabilities to a machine and letting it interact with centralized information or computing resources. At the early stages this will be possible only on a small scale, and only for machines for which the marginal cost of bundling a mobile terminal does not add much to the final price of the device. Therefore, such services may be confined to specialized devices. Later, as the unit terminal cost drops, this may become more common for generic consumer devices.

Another example, telemetry applications, involves remote monitoring of devices and events. This was one of the most commonly quoted applications—although perhaps not as commonly used—in the early wireless data systems. With the advent of location-based services, new types of fleet management applications can be expected to appear as an extension and enhancement for the existing telemetry services. This might require the integration of the application in corporate network environments, thus associating the location-based application with a community-based service where the community member is a vehicle belonging to the fleet, all supported over an MVPN associated with the company that owns the fleet. Possible customers for such services include transportation companies and radio taxi services to convert to cellular-data-systems-based car booking and assignment (although at the end the human will use information exchanged by the in-car computer system and the centralized reservation system, there is a high degree of human independence in this service). Other applications may include the backup line provisioning for mission-critical environments in case of failure of a wireline link on some access routers, as well as remote monitoring.

# Mobile Virtual Network Operator

The use of Mobile Virtual Network Operators is a concept until recently not commonly known by nonspecialists, and even MVNO customers may potentially be unaware of who is providing the service and with what technology. From a customer point of view, there is no difference between the services provided by an MVNO versus a traditional wireless carrier, while in reality the services are provided via some form of agreement between the MVNO and the actual cellular networks operator. With the advent of alternate wireless access network providers, such as WLAN Internet service providers, or WISPs, MVNOs may also offer service via a variety of access media. In addition, there may be different flavors of MVNO depending on its level of engagement in the physical network operations. Lightweight MVNO differentiates based on unique customer care and branding, while full-scale MVNO not only focuses on branding but also operates some critical network elements.

## Lightweight MVNO

A *lightweight MVNO* focuses its activity on mobile customer acquisition and retention, marketing campaigns, customer care, and branding. These operators do not want to enter into the technical aspects of managing services, and they believe services offered via the network of a partner, or multiple partners, are sufficient to satisfy their customer needs and to prevent them from switching to the competition. Typically, they lease space on existing carrier HLR and AAA servers and use that carrier's infrastructure or roaming agreements. Alternatively, MVNOs might use their own HLRs and AAA servers and simply have roaming agreements in place with all the national carriers with special conditions regulated by their MVNO status.

These operators do not require huge up-front investments, and they can invest all their efforts in subscribers' growth and satisfaction, and can leverage their branding power by letting existing players bid for their business. One example of an existing Voice MVNO is Virgin Mobile in the United Kingdom and the United States, where the strong Virgin brand and distribution infrastructure allowed it to sign up millions of subscribers in a relatively short time period. The MVNOs may also define their own applications partners in order to further differentiate themselves from the competition and the services provided by a wireless carrier whose infrastructure they are using. The MVNOs are normally expected to operate in the consumer market space, although they may later branch out to serve small businesses.

## Full-Scale MVNO

A *full-scale MVNO* adds value to the plain access network connectivity offered by incumbent operators by operating some elements in the core network. In a 3GPP-defined system, for instance, the MVNO may want to operate the HLR, the AAA servers, and the GGSN or logical partitions of GGSNs (virtual GGSNs) offered to them by existing wireless carriers for sharing. In 3GPP2-defined systems they may operate parts of the clusters of Home Agents, PDSN, and AAA servers. Full-scale MVNOs may operate applications jointly with existing ASPs and use these and other similar relationships to address the corporate market via managed VPN services and corporate data services applications. A traditional wireline carrier serving significant numbers of business customers may decide to add wireless access support for them to extend its product line and act as MVNO, thus retaining customers' ownership. This class of MVNOs is willing to compete for technical elements of service delivery, either because they believe their experience gives them a competitive advantage or because they know how to handle the existing customer base and they believe it is in their own best interest not to lose it to competitive offers from wireless carriers.

## MVPN in an MVNO Environment

It is clear that the existence of MVNOs needs to be taken into account when MVPN's technical solutions are defined, since the MVNO itself may be a special kind of wireless carrier customer. As mentioned, the MVNO sometimes does not want to engage in network operations, and the wireless carrier therefore offers complete outsourcing of network operations for them. In such lightweight MVNO cases, the MVPN technical solutions do not change much.

The MVNO may ask the wireless carrier to offer an MVPN service over which the MVNO implements a captive portal (see Chapter 5) where their subscribers are being authenticated and where they select the services they want to subscribe to or want to use during a particular session. This VPN may provide the access to multiple areas associated with bundles of such services. The MVNO may ask the wireless carrier to construct other additional VPNs to support application-specific networks that only subscribers to these applications have the right to access. Additionally, if the lightweight MVNO had to support corporate networks, they would ask the partner to entirely manage the technical aspects of the SLA and service integration, perhaps together with a third-party technical partner who may have a relationship with a number of enterprise customers the MVNO could leverage.

Alternatively, the MVNO could lease GGSNs or virtual GGSNs (or HAs in the 3GPP2 systems) from the wireless carrier and manage them according to the rules set forth in the contract defining this business relationship. It also may fully manage the integration services and corporate customer support. In this case, the ownership of the gateway node implies that the MVNO may freely manage the VPNS they offer to their customers, and the MVNO must also operate the DNS server necessary for networks to resolve the APNs to IP addresses on the GGSNs it operates.

In yet another scenario, the MVNO might offer voluntary MVPN services based on the IPSec or Mobile IP VPN client to their customers, to provide for the support of multiple access technologies and to terminate the IPSec connection at a VPN gateway or HA residing in the MVNO data center (or let the customers themselves own or lease the CPE equipment), before delivering traffic to the appropriate customer network. This scenario makes use of platforms specialized to handle a large number of IPSec tunnels and to support virtual VPN GW service so that each customer network is assigned a virtual VPN GW, or to apply switching between ingress and egress virtual interface based on the customer network identity and not based on IP routing. However, we do not believe this will be a mainstream scenario because of heterogeneous requirements from corporations and the potential overhead of the air interface, especially when used with legacy air interface technologies such as GSM.

## WLAN/Cellular Convergence and MVPN

In this section we consider the main mechanisms used to create converged MVPNs integrating wireless packet data technologies and systems such as Wireless LAN, GPRS, UMTS, and CDMA2000. Convergence of Wireless LAN hot spots with the cellular systems is considered here for a few reasons. First, it well exemplifies the problems and benefits associated with integration of different wireless networks in general, especially with regard to private networking over the resulting combined system. Second, it has a strong business case, further substantiated by the growing interest by customers, wireless carriers, and such new entrants in the world of mobile data services as local coffee shops, small businesses, hotels, airports, and railway stations. Finally, since different cellular systems support different compulsory MVPN mechanisms, it is easier to examine their properties and then apply the observation to the real-world design of the converged WLAN/single cellular system combination.

## WLAN and Cellular Integration

At first look, the integration of WLAN and a typical cellular system seems unlikely. After all, WLAN, as we discussed in Chapter 3, is not a standardized system; it doesn't include subscriber data handling, location management, or mobility management, nor does it include built-in support for macro-mobility, fast handover, roaming, user authentication, and other attributes traditionally associated with a telecommunications system such as GSM. On the other hand, from both residential user and business application perspective the integration of WLAN and cellular systems makes sense. If properly integrated, WLAN can complement a typical 2G or 3G cellular system in many ways. WLAN throughput rates are far superior to those of even the latest wireless packet data systems such as UMTS. In addition, WLAN equipment is significantly less expensive and easier to install and support. While satisfying bandwidth-hungry road warriors, WLAN can at the same time offload the cellular infrastructure by serving subscribers in highly congested areas. WLAN combined with cellular systems can serve as a foundation for new revenue-generating data services such as high-speed MVPN access and other wireless data services requiring high throughput bearer.

These factors recently spurred a new breed of unlicensed wireless LAN operators known as Wireless Internet service providers, who install WLANs in strategic locations, where untethered high-speed data access is likely to be required. These so-called *hot-spot* locations, or WLAN *hot spots*, are fast gaining popularity with mobile data users and can be formed, upon mergers and acquisitions or federations of Wireless ISPs, into a network of locations capable of covering a wide area, especially when combined with a well-developed packet-data-based cellular system.

## WLAN Integration Methods

WLAN/cellular integration is a relatively recent phenomenon, but there are already a number of mechanisms available that would allow for more or less seamless integration between the two. 3GPP SA1 has recently put a requirement for the interworking of WLAN with UMTS networks and to generate the standards that would govern it. As a consequence, 3GPP SA2 is currently working on defining the architecture and system aspects of WLAN and cellular systems integration. Another group, ETSI BRAN (for Broadband Radio Access Networks), was tasked with specifying mechanisms to define HiperLAN/2. Their resulting interim work can be seen in [ETSI BRAN TR101.957]. This document also attempts to classify WLAN/GPRS integration methods by defining two integration options: loose and tight (see Figure 9.4).

**Loose Internetworking**

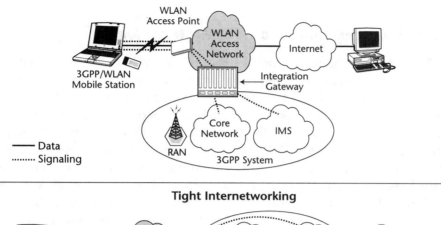

**Tight Internetworking**

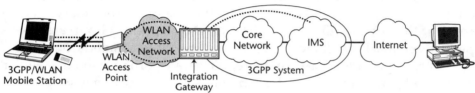

**Figure 9.4**   Tight and loose cellular/WLAN integration options.

**NOTE** HiperLAN/2 is a system developed by the ETSI in 5-GHz band. It is almost identical to 802.11a at the physical layer, where it uses orthogonal frequency-division multiplexing (OFDM) and has the same data rates.

*Loose* integration refers to a scenario where the WLAN traffic does not traverse the GPRS core and instead is directly routed to the Internet, so the only function shared by the two systems is authentication and accounting. *Tight* integration refers to a scenario when the WLAN traffic is handled by the elements in the GPRS core network or UMTS PS domain, such as GGSN, SGSN, and Charging Gateway Function. This classification is only applicable in the GPRS and UMTS systems framework, so instead we will analyze the WLAN/cellular integration methods based in part on the authentication mechanisms, paying special attention to IMSI- and NAI-based authentication with Mobile IP and support for MVPN and other services targeting the end-user.

Integration may allow for the same wireless data services across the different access technologies, as well as delivery of the same Mobile VPN services (in tight integration scenarios by using the same APN on the GGSNs used by the subscriber in the cellular network; or in loose integration scenarios, by using Mobile IP-based network access or voluntary tunneling-based solutions).

### IMSI-Based Authentication for WLAN Integration

The IMSI-based mechanism assumes that the WLAN user device can be authenticated by the common cellular systems infrastructure the same way any cellular device is authenticated. For example, in GPRS the MS is authenticated via a built-in Subscriber Identity Module chipcard, so the first requirement for WLAN/GPRS IMSI-based authentication is the support of a SIM reader in the WLAN mobile terminal equipment. A good example is the WLAN PCMCIA card with a built-in SIM reader manufactured by Nokia, which introduced a combined SIM-based GPRS/WLAN device that addressed many user-side integration and compatibility issues. Today, over the radio link, this requires the use of a proprietary mechanism. For this reason, IEEE or the IETF's PANA Working Group are developing an approach to achieve this same functionality in standard fashion.

Network support for this scenario requires transporting standard cellular system authentication material from the WLAN access gateway to an interim gateway where the subscriber authentication material is downloaded from the HLR. This can be addressed by the introduction of a new network element that would integrate the WLAN and GPRS infrastructures and bridge the AAA protocol used by WLAN, such as RADIUS, with the MAP or TIA [IS41] signaling and AAA protocols used in cellular networks. For simplicity we will refer to such a device as an *integration gateway*. This gateway, depicted in Figure 9.4, may include other functionalities depending on the degree of WLAN integration with the cellular infrastructure.

In a tightly coupled integration scenario, the integration gateway can act in 3GPP systems as an SGSN from the cellular system point of view, interfacing to the CGF and GGSN and to the SGSN serving cellular radio access network. The compulsory MVPN access in this case will be supported by the GPRS infrastructure in a manner similar to that used for GPRS terminal users (see Chapter 6). In an alternative architectural approach—that is, loose integration—the integration gateway can be implemented as standalone equipment. In this case, only the WLAN AAA protocol and MAP or TIA [IS41] signaling and AAA protocols used in cellular networks are interworked, and traffic is directly delivered to the Internet. This approach is preferable and is fast gaining both vendor support and market acceptance.

The main function of the integration gateway in combined WLAN/GPRS systems is to convert the WLAN RADIUS-based accounting data into a

GPRS billing format (typically in the form of G-CDRs files complying with 3GPP requirements [TS32.015]) and interworking RADIUS authentication with MAP-based access to the information stored in the HLR. The GPRS billing system may be upgraded to be able to identify the source of the CDRs in order to accurately bill the customer for WLAN versus GPRS usage.

IMSI-based authentication is not directly applicable in the CDMA2000 environment, since CDMA2000 authentication is not based on a SIM card or its equivalent.[2] Currently, the CDMA2000 authentication and accounting done by the PDSN and the HA (in the case of Mobile IP) is based on RADIUS. Because RADIUS authentication is also used in WLAN, the task of WLAN integration is relatively straightforward. The only requirement in this case would be to convert the WLAN accounting parameters to the form specified for CDMA2000 systems by TIA [IS835]. In CDMA2000 these consist of radio-specific parameters collected by the RAN and core-network-specific parameters collected by the PDSN. The PDSN then forms a Usage Data Record (UDR) consisting of both parameters, which is forwarded to a local AAA Server and possibly communicated over the AAA broker infrastructure (as discussed in Chapter 7).

### NAI-Based Authentication and Mobile IP

Recall from the previous chapters that CDMA2000 core data networking is based on Simple IP and Mobile IP access methods, both of which rely on IETF-developed standards and GPRS and UMTS systems also allow for a Mobile IP-based private network access option. Also recall that Mobile IP can be supported over virtually any access network technology with no changes to the infrastructure and the protocol itself. These factors have led many in the industry to combine this technology with the equally ubiquitous NAI-based authentication, thereby introducing another WLAN/cellular integration method. In contrast to the IMSI-based approach, this method more easily integrates with existing ISP and corporate AAA environments—although it requires the cellular operator in GPRS/UMTS systems to operate a RADIUS-based AAA server compatible with the WLAN AAA protocol.

There are two possible deployment scenarios for NAI-based authentication systems:

---

[2] This situation, however, may soon change with SIM-based authentication in CDMA2000 systems being developed by 3GPP2.

- The MS is not allowed to hand off to the cellular system, and vice versa, without losing the session continuity. Therefore, the integration is only at the AAA level, and user mobility is expected to be limited if it exists at all. However, this model does not appear to be suffering much from mobility limitations, since in most of the cases, WLAN users are using laptops in hot spots and are therefore fairly static.

- The intersystem handoff is enabled by Mobile IP support in the cellular infrastructure and in the WLAN infrastructure. Mobile IP-based WLAN/cellular infrastructure would require the Mobile IP FA to be supported by the WLAN access point or at intermediate gateways serving a group of APs (see Figure 9.5).

In the case of the CDMA2000, when Mobile IP access method is used, Mobile IP is of course natively supported. Users equipped with laptops that support WLAN and CDMA2000 access—for example, via two PCM-CIA cards—can roam almost seamlessly between two different types of wireless access technologies, while preserving the user's IP address and end-to-end communications including voluntary tunneling or TLS session [RFC2246] to a private network. NAI-based authentication has the advantage of being supported by many standard types of WLAN equipment and standard IP-based protocols and hence by client software. Standard RADIUS proxy procedures, broker infrastructures, and IP authentication parameters are used for roaming and authentication.

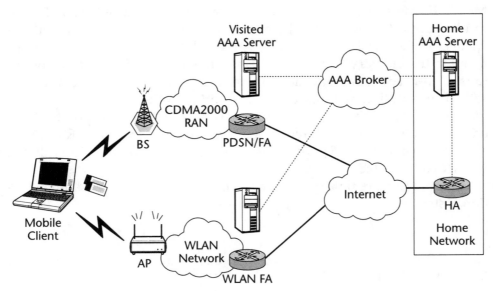

**Figure 9.5**   Mobile IP-based WLAN/cellular integration.

In the case of UMTS and GPRS, a cellular operator offering Mobile IP-based WLAN integration is required to deploy a GGSN integrating Mobile IP FA functionality. This requires configuring the terminal to request the Mobile IP-capable APNs supported by these GGSNs, and to store subscription information in the HLR that allows the user to access these APNs.

**NOTE** Using Mobile IP bypasses the need of a standardized protocol for carrying authentication for network access (as standardized by the IETF PANA WG), since this functionality is provided by Mobile IP itself. This may be yet another advantage of deploying a Mobile IP-based solution in the short term.

## Summary

In this chapter, we summarized some trends in the wireless data industry and tried to extrapolate how they will influence service delivery to mobile subscribers. We discussed the kinds of wireless services and applications we may find in future wireless networks and some possible service scenarios, including those derived from the integration of WLAN with the cellular infrastructures.

This chapter also concludes this book, which we hope has helped you acquire an understanding of many aspects of provisioning and architecting MVPN and other advanced data services in the mobile environment, as well as a general understanding of the data mobility and wireless services industry. We also hope that this book will constitute only the beginning of your research on the subject and will encourage you to investigate other areas of mobile service delivery and wireless data networking.

# Mobile IP Extensions

This appendix describes extensions that have been added to the original Mobile IP protocol to address some of its shortcomings in meeting commercial service requirements such as AAA support and dynamic home address assignment, and to facilitate its use in cellular systems such as CDMA2000.

## Challenge/Response Extensions

Mobile IP Challenge/Response extensions described in [RFC3012] are extensions to Mobile IP agent advertisements and registration requests defined in Chapter 2. They are designed to address some of the Mobile IP security problems, such as FA vulnerability to replay-based attacks.

The *Challenge* extension format is shown in Figure A.1. It is added to the *agent advertisement* message sent by an FA so that MNs that receive it can create a valid registration request message (that is, with the latest valid challenge value). This process allows it to be positively authenticated.

```
0                   1                   2
0 1 2 3 4 5 6 7 8 9 0 1 2 3 4 5 6 7 8 9 0 1 2 3 4 5 6 7 8 9 0 1
```

| Type = 24 | Length | Challenge |
|-----------|--------|-----------|

**Figure A.1**   Mobile IP Challenge extension format.

The MN-FA Challenge extension, shown in Figure A.2, is an extension to the MN *registration request* message that is issued in response to a challenge in an agent advertisement. The MN-FA Challenge includes the latest challenge value (identical to the one in the Challenge extension), indicating that this MN is not responding for a previously issued FA advertisement. Note that MN-FA Challenge extension is mandatory in CDMA2000 environment as per [IS835] and must be included in every agent advertisement.

[RFC3012] also defines the *Generalized Authentication extension* format designed to accommodate control messages for future Mobile IP extensions that may be used to exchange authentication information between the MN and other network elements such as AAA servers. Generalized Authentication extensions, depicted in Figure A.3, specify a new application type with subtypes into which the new authentication applications can be classified. In this figure the security parameter index (SPI) indexes the SA table.

For example, subtype 1 of the Generalized Authentication extension, also defined in [RFC3012], has been assigned to the MN-AAA authentication extension and may be used in place of (or together with) the Mobile-Foreign Authentication extension defined in the original [RFC2002] when the Challenge extension is included in FA advertisement message. Like the MN-FA the extension, MN-AAA extension is also mandatory for CDMA2000 systems compliant with [IS835].

Whenever the Challenge extension is used, the MS must include the latest challenge value in the MN-FA Challenge extension to its registration request whenever a security association with the FA is not pre-established. In this case MN-AAA and, optionally, NAI extensions (discussed in the next section) must also be included in the RRQ. The challenge value is optional, however, when the MN-FA security association exists.

The computation of the authenticator field of the MN-AAA extension is defined by the SPI value. Refer to [RFC3012] for further details.

```
0                   1                   2
0 1 2 3 4 5 6 7 8 9 0 1 2 3 4 5 6 7 8 9 0 1 2 3 4 5 6 7 8 9 0 1
```

| Type = 132 | Length | Challenge Value |
|------------|--------|-----------------|

**Figure A.2**   Mobile IP MN-FA Challenge extension format.

```
0                   1                   2
0 1 2 3 4 5 6 7 8 9 0 1 2 3 4 5 6 7 8 9 0 1 2 3 4 5 6 7 8 9 0 1
```

| Type | Subtype | Length |
| --- | --- | --- |
| Security Parameter Index (SPI) | | |
| Authenticator | | |

**Figure A.3**   Mobile IP Generalized Authentication extension format.

## NAI Extension

The Network Access Identifier (NAI) extension, also referred to as the MN-NAI extension, provides a way for an MN to identify itself to a foreign network, along with the identity verification infrastructure which can be queried to authenticate the user—and possibly obtain other information useful to set up the session, such as security keys and QoS or service charging parameters. To achieve this, the NAI extension is included in the MS's registration request. The NAI itself, carried in the NAI extension, defines a format for the user identity (described in [RFC2486]) in the form *user@domain*. The *user* component provides a user identity and the *domain* component the identity of a verification infrastructure that can be used to verify the user identity. The user identity must be unique within the "domain." The NAI extension is described in [RFC2794], which obsoletes [RFC2290]. Its format is shown in Figure A.4.

If the NAI is used to identify the MS, then the foreign network does not require its home address, so the MS without a home address can still be positively authenticated by the network. In this case the Home Address field in the registration request message is set to 0, which would signal the need to assign a home address to the MS. The home address is then included in the registration reply from the FA or HA. As such, the NAI extension allows not only for roaming support but also for the support of dynamic home address assignment. The NAI is included in the registration request before the MN-HA and MN-FA extensions.

```
0                   1                   2
0 1 2 3 4 5 6 7 8 9 0 1 2 3 4 5 6 7 8 9 0 1 2 3 4 5 6 7 8 9 0 1
```

| Type | Length | MN-NAI |
| --- | --- | --- |

**Figure A.4**   NAI extension format.

**NOTE** The application of these extensions to CDMA2000 Mobile IP-based networking is governed by [IS835].

## Private Extensions

Vendors or operators often require that proprietary extensions to a protocol be defined to provide unique features or simply for diagnostic or versioning reasons. For this reason, Mobile IP vendor-/organization-specific extensions have been defined in [RFC3115]. There are two types of vendor-specific extensions:

- Critical (CVSE)
- Normal (NVSE) vendor-/organization-specific extensions

One basic difference between critical and normal extensions is that when the critical extension is encountered but not recognized, the message containing the extension *must* be silently discarded. On the other hand, when a normal, vendor-/organization-specific extension is encountered but not recognized, the extension *should* be ignored, but the rest of the extensions and message data *must* still be processed. Another difference is that a critical, vendor/organization extension has a length field of 2 octets, whereas the NVSE has a length field of only 1 octet. Figures A.5 and A.6 describe the CVSE and NVSE, respectively.

```
 0                   1                   2                   3
 0 1 2 3 4 5 6 7 8 9 0 1 2 3 4 5 6 7 8 9 0 1 2 3 4 5 6 7 8 9 0 1
+-+-+-+-+-+-+-+-+-+-+-+-+-+-+-+-+-+-+-+-+-+-+-+-+-+-+-+-+-+-+-+-+
|     Type      |   Reserved    |            Length             |
+-+-+-+-+-+-+-+-+-+-+-+-+-+-+-+-+-+-+-+-+-+-+-+-+-+-+-+-+-+-+-+-+
|                          Vendor/Org-ID                        |
+-+-+-+-+-+-+-+-+-+-+-+-+-+-+-+-+-+-+-+-+-+-+-+-+-+-+-+-+-+-+-+-+
|        Vendor-CVSE-Type       |     Vendor-CVSE-Value ...     |
+-+-+-+-+-+-+-+-+-+-+-+-+-+-+-+-+-+-+-+-+-+-+-+-+-+-+-+-+-+-+-+-+
```

**Figure A.5**   Critical vendor/organization extensions.

```
 0                   1                   2                   3
 0 1 2 3 4 5 6 7 8 9 0 1 2 3 4 5 6 7 8 9 0 1 2 3 4 5 6 7 8 9 0 1
+-+-+-+-+-+-+-+-+-+-+-+-+-+-+-+-+-+-+-+-+-+-+-+-+-+-+-+-+-+-+-+-+
|     Type      |    Length     |            Reserved           |
+-+-+-+-+-+-+-+-+-+-+-+-+-+-+-+-+-+-+-+-+-+-+-+-+-+-+-+-+-+-+-+-+
|                          Vendor/Org-ID                        |
+-+-+-+-+-+-+-+-+-+-+-+-+-+-+-+-+-+-+-+-+-+-+-+-+-+-+-+-+-+-+-+-+
|        Vendor-NVSE-Type       |     Vendor-NVSE-Value ...     |
+-+-+-+-+-+-+-+-+-+-+-+-+-+-+-+-+-+-+-+-+-+-+-+-+-+-+-+-+-+-+-+-+
```

**Figure A.6** Normal vendor/organization extensions.

# CDMA2000 RADIUS Accounting Attributes

This appendix describes the formats for RADIUS attributes defined by 3GPP2 and standardized in [IS835]. The generalized format shared by all RADIUS attributes is shown in Figure B.1.

Currently 3GPP2 RADIUS attributes implemented by major vendors include the following:

- Accounting container
- DiffServ option
- IKE attributes such as pre-shared secret, pre-shared secret request, and KeyID
- HA attribute
- Security level
- Reverse tunnel specification

```
0               1               2               3
0 1 2 3 4 5 6 7 0 1 2 3 4 5 6 7 0 1 2 3 4 5 6 7 0 1 2 3 4 5 6 7
```

| Type = 26 | Length>9 | Vendor ID | |
|---|---|---|---|
| Vendor ID | | Vendor Type | Vendor Length |
| Vendor Value | | | |

**Figure B.1**   3GPP2 RADIUS attribute format.

Reproduced under written permission from Telecommunications Industry Association.

# Accounting Container

Some fields in RADIUS accounting records might change in the middle of the session. IS835 provides for two methods to send changes in parameters from PDSN to RADIUS servers. One of them is to send a "stop" followed by a "start" record, with the changed values to stop the current accounting record and initiate a new one in the event of a parameter change.

Another method utilizes "accounting containers" to store and forward information in the format depicted in Figure B.2. The changed fields of the accounting record and change reasons such as tariff boundary, handoff, and parameter change are stored in the appropriate fields in the container. This technique allows the RADIUS server to continue to store accounting information without interruption.

## IKE Attributes

IKE attributes include pre-shared secret requests indicating if a pre-shared secret required during PDSN and HA negotiations, a KeyID attribute sent by the home RADIUS server to the PDSN, and the "S" attribute used during generation of pre-shared secret sent by the home RADIUS server to the HA during PDSN-to-HA IKE negotiation. Table B.1 summarizes IKE attributes parameters.

```
0               1               2               3
0 1 2 3 4 5 6 7 0 1 2 3 4 5 6 7 0 1 2 3 4 5 6 7 0 1 2 3 4 5 6 7
```

| Type = 26 | Length>14 | Vendor ID = 5535 | |
|---|---|---|---|
| Vendor ID (cont) | | Vendor Type | Vendor Length |
| Container Reason | | Event Timestamp | |
| Event Timestamp (cont) | | 3GPP2 AVPs | |

**Figure B.2**   Accounting container format.

Reproduced under written permission from Telecommunications Industry Association.

**Table B.1** IKE Attributes Parameters

| ATTRIBUTE | VENDOR TYPE | VENDOR LENGTH | VENDOR VALUE |
|---|---|---|---|
| "S" request | 13 | 6 | 0 = No request for S from HA<br>1 = Request for S from HA |
| "S" lifetime | 12 | 6 | Number of seconds since 1/1/1970 00:00 UTC |
| "S" | 11 | 3 | Value of the secret |
| KeyID | 8 | 22 | HAAA address + FAAA address + timestamp[1] |
| Pre-shared secret request | 1 | 6 | 1 = Requested by PDSN<br>0 = Not requested by PDSN |
| Pre-shared secret | 3 | 18 | Secret key value |

[1] The event timestamp contains information about the beginning and ending of an accounting session. Its value field contains the number of seconds since 1/1/1970 00:00 UTC (similar to "S" lifetime).

## Security Level, HA, Reverse Tunnel, and DiffServ Attributes

All of these attributes are optionally included in Access-Accept messages. The parameters of these attributes are included in Table B.2:

- *Security level* is an optional attribute sent from the home to the visited AAA server, indicating the type of security the visited network must provide to the MN.

- *Reverse tunnel* is an attribute indicating whether a reverse tunnel must be created between visited and home networks.

- The *HA* attribute carries an HA address.

- The *DiffServ* attribute is used by RADIUS servers to define the use of DiffServ (described in Chapter 2) to provide quality of service to the data traffic passing through the PDSN. The values of this attribute are specified according to [RFC2597] and [RFC2598].

**Table B.2**   IKE Attributes Parameters

| ATTRIBUTE | VENDOR TYPE | VENDOR LENGTH | VENDOR VALUE |
|-----------|-------------|---------------|--------------|
| Security Level | 2 | 6 | 1 - IPSec required for registration messages<br>2 - IPSec required for tunnels<br>3 - IPSec required for tunnels and messages<br>4 - IPSec not required |
| Reverse tunnel | 4 | 6 | 0 = Not required<br>1 = Required |
| HA | 7 | 6 | IP address |
| DiffServ | 5 | 6 | Set according to RFC2597 and RFC2598 |

# RADIUS Usage in 3GPP

In this appendix we provide some details on how RADIUS is typically used in 3GPP systems. 3GPP defines RADIUS usage for two purposes: classic network access AAA functionality and interaction with application servers (for instance, WAP gateways). The first role is well defined and directly derived from the common way RADIUS is used in remote access servers. The second role was instead traditionally left to proprietary solutions each vendor decided as most suitable for its needs.

Recognizing this was a limitation for multivendor solutions and procurement, the industry has initiated a strong push in 3GPP toward the standardization of the usage of RADIUS (and in particular RADIUS accounting) in interacting with application servers The set of proprietary RADIUS attributes that classically had been used to exchange information with application servers has been replaced by an agreed-upon set of 3GPP vendor-specific attributes (VSAs). Also, the formatting and usage of RADIUS messages and the criteria of inclusion of attributes has been agreed upon, as well as the conditions under which user traffic can be forwarded in relation to the outcome of RADIUS authentication and accounting procedures. All this has been defined as an extension of the already existing [3GPP TS29.061] specification and as a change to [GSM TS09.61].

While it is clear why in a MVPN environment RADIUS AAA usage for network access is required, it may not be so clear why we should address in this book how RADIUS is used for interacting with applications servers. For one, in real deployments it may happen that the two different usages are not separated; rather, the GGSN interacts with a single RADIUS server that is configured and programmed to interact, as a proxy, with application servers and AAA servers separately, perhaps after some RADIUS attributes preprocessing.

On the other hand, in some instances MVPN enables its members to use a set of applications. These applications may benefit from some information delivered by the GGSN to the associated applications servers via a RADIUS interface. As such, it is extremely useful to understand the fundamentals on how to use the 3GPP RADIUS interface to application servers and how to enable the GGSN to perform this essential functionality.

## Possible Network Configurations

At least in theory, there are multiple possible network configuration scenarios. In one scenario, the GGSN interfaces with a single AAA server, which then is programmed to proxy toward RADIUS authentication servers, accounting servers, and application servers. This offloads to the AAA proxy server the complexity of interacting with all the other subsystems, but it also adds flexibility, since the AAA proxy server may be a programmable platform, with much more storage and cheaper computing power than the GGSN.

In another scenario, the GGSN directly interfaces to one authentication server, one accounting server, and one application server. This adds load to the GGSN in that it has to interact with multiple entities and keep state and resources adequate for this purpose. In another scenario, the GGSN interfaces only to an application server, if the wireless access authentication is sufficient (i.e., IMSI-based authentication taking place at network attach time) and no extra level of authentication is necessary for gaining access to the network where the application server is. This is typically the case with WAP applications, where time-based charging was used or data-volume usage data was collected via the Ga interface.

Yet one can imagine having the GGSN interact with an authentication server as well, if an additional level of authentication is deemed necessary. Note, however, that the use of RADIUS accounting to deliver volume or time-based usage is not strictly necessary if the information contained in GTP'-based [3GPP TS32.015] accounting is used. However, RADIUS-based accounting may prove to be necessary when the network where application servers are hosted belongs to a third party that requires usage data be delivered in RADIUS format. Because of all these possible scenarios, a GGSN should come with the ability to configure separate accounting and authentication servers, and possibly an additional RADIUS accounting interface to an application server.

## RADIUS for Authentication

The usage of RADIUS for network access authentication follows the common practice in the remote access servers industry. However, a few remarks are in order. It has become common practice to include the MSIDSN of the MS in the CALLING-STATION-ID attribute, and the value of the APN in the CALLED STATION-ID attribute.

The use of ACCESS-CHALLENGE message has some limitations. If the PPP PDP type is used, this message can be used in a normal manner. However, when the message is received for an APN in *IP with Protocol Configuration Options* access mode (see Chapter 6), then it is not possible for the system to challenge the MS and for the MS to return a challenge response. Therefore, the message is interpreted as an Access Reject. To avoid undesirable disconnection events, it may be wise not to enable the AAA server to issue challenge messages.

When using RADIUS authentication in a 3GPP system, you should consult the 3GPP specification [TS29.061] to gather information on how to best configure and use RADIUS authentication in GPRS/UMTS. The specification is extremely helpful for using the RADIUS authentication features appropriately.

## RADIUS for Accounting

The usage of RADIUS for accounting in 3GPP-defined systems is pretty much equivalent to the usage of RADIUS accounting in wireline networks, with the exception that there are some specific applications of RADIUS accounting that are particularly important in wireless networks. For instance, RADIUS interim accounting may be used to implement prepaid services (by triggering the sending of RADIUS Accounting Request interim updates based on the volume of time-based thresholds), as well as tariff time plans (by triggering the sending of RADIUS Accounting Request interim updates based on specific time of day or day of week).

This usage of RADIUS accounting may be adopted in wireline networks too, but it is particularly useful to adapt the service paradigm to the current prepaid service paradigm many GSM users are experiencing. When a user is roaming, the accounting server may record this based either on the 3GPP-SGSN-IP-ADDRESS VSA or the 3GPP-GGSN-MCC-MNC VSA (when a GGSN in the visited network was used) described in [3GPP TS29.061]. Note that in UMTS you can associate multiple bearers to a single session. As such, a session is up when one or more of these bearers are active. This requires the GGSN to indicate when the last existing bearer of a session is torn down, in order to enable, for instance, the release of the IP address assigned via RADIUS. 3GPP also defines the optional use of a uncommonly used DISCONNECT message [RFC2822]. This may be used to disconnect users exhausting their prepaid balance.

# RADIUS for Interaction with Application Servers

As mentioned, RADIUS plays an important role in advanced wireless data services provisioning because it relays session-related information to application servers To this effect, [3GPP TS29.061] defines a set of 3GPP vendor-specific attributes that can be used to transfer session-related parameters such as the QoS value, the SGSN IP address, the GPRS charging identifier, and the IMSI, via RADIUS Accounting. See Table C.1 for a detailed listing of these attributes.

The specification also defines what format of RADIUS messages should be used (that is, which attributes to include in a message and under which conditions must each attribute be contained in a message) for a 3GPP-compliant system. In particular, the Accounting Request START message must include the MS-ISDN in the calling station ID attribute and the APN in the called station ID in order to map a subscriber identity to the IP address it uses, carried in the framed protocol address attribute.

**Table C.1**  Sub-attributes of the 3GPP Vendor-Specific Attribute (from 3GPP 29.061 v5.2.1)

| SUB-ATTR # | SUB-ATTRIBUTE NAME | DESCRIPTION | PRESENCE REQUIREMENT | ASSOCIATED ATTRIBUTE (LOCATION OF SUB-ATTR) |
|---|---|---|---|---|
| 1 | 3GPP-IMSI | IMSI for this user | Optional | Access Request, Accounting Request START, Accounting Request STOP, Accounting Request interim update |
| 2 | 3GPP-Charging-ID | Charging ID for this PDP context (this, together with the GGSN address, constitutes a unique identifier for the PDP context) | Optional | Access Request, Accounting Request START, Accounting Request STOP, Accounting Request interim update |
| 3 | 3GPP-PDP Type | Type of PDP context, e.g., IP or PPP | Conditional (mandatory if attribute 7 is present) | Access Request, Accounting Request START, Accounting Request STOP, Accounting Request interim update |
| 4 | 3GPP-CG-Address | Charging gateway IP address | Optional | Access Request, Accounting Request START, Accounting Request STOP, Accounting Request interim update |
| 5 | 3GPP-GPRS-Negotiated-QoS-Profile | QoS profile applied by GGSN | Optional | Access Request, Accounting Request START, Accounting Request STOP, Accounting Request interim update |
| 6 | 3GPP-SGSN-Address | SGSN IP address that is used by the GTP control plane for the handling of control messages. It may be used to identify the PLMN to which the user is attached. | Optional | Access Request, Accounting Request START, Accounting Request STOP, Accounting Request interim update |

*(continues)*

**Table C.1**  Sub-attributes of the 3GPP Vendor-Specific Attribute (from 3GPP 29.061 v5.2.1) *(Continued)*

| SUB-ATTR # | SUB-ATTRIBUTE NAME | DESCRIPTION | PRESENCE REQUIREMENT | ASSOCIATED ATTRIBUTE (LOCATION OF SUB-ATTR) |
|---|---|---|---|---|
| 7 | 3GPP-GGSN-Address | GGSN IP address that is used by the GTP control plane for the context establishment. It is the same as the GGSN IP address used in the GCDRs. | Optional | Access Request, Accounting Request START, Accounting Request STOP, Accounting Request interim update |
| 8 | 3GPP-IMSI-MCC-MNC | MCC and MNC extracted from the user's IMSI (first 5 or 6 digits, as applicable from the presented IMSI). | Optional | Access Request, Accounting Request START, Accounting Request STOP, Accounting Request interim update |
| 9 | 3GPP-GGSN-MCC-MNC | MCC-MNC of the network the GGSN belongs to. | Optional | Access Request, Accounting Request START, Accounting Request STOP, Accounting Request interim update |
| 10 | 3GPP-NSAPI | Identifies a particular PDP context for the associated PDN and MSISDN/IMSI from creation to deletion. | Optional | Access Request, Accounting Request START, Accounting Request STOP Accounting Request interim update |
| 11 | 3GPP- Session-Stop-Indicator | Indicates to the AAA server that the last PDP context of a session is released and the PDP session has been terminated. | Optional | Accounting Request STOP |

**Table C.1** Sub-attributes of the 3GPP Vendor-Specific Attribute (from 3GPP 29.061 v5.2.1) (*Continued*)

| SUB-ATTR # | SUB-ATTRIBUTE NAME | DESCRIPTION | PRESENCE REQUIREMENT | ASSOCIATED ATTRIBUTE (LOCATION OF SUB-ATTR) |
|---|---|---|---|---|
| 12 | 3GPP- Selection-Mode | Contains the selection mode for this PDP context received in the Create PDP Context request message. | Optional | Access Request, Accounting Request START, Accounting Request STOP, Accounting Request interim update |
| 13 | 3GPP-Charging-Characteristics | Contains the charging characteristics for this PDP context received in the Create PDP Context request message (only available in R99 and later releases). | Optional | Access Request, Accounting Request START, Accounting Request STOP, Accounting Request interim update |
| 14 | 3GPP-CG-IPv6-Address | Charging gateway IPv6 address | Optional | Access Request, Accounting Request START, Accounting Request STOP, Accounting Request interim update |
| 15 | 3GPP-SGSN-IPv6-Address | SGSN IPv6 address that is used by the GTP control plane for the handling of control messages. It may be used to identify the PLMN to which the user is attached. | Optional | Access Request, Accounting Request START, Accounting Request STOP, Accounting Request interim update |
| 16 | 3GPP-GGSN-IPv6-Address | GGSN IPv6 address that is used by the GTP control plane for the context establishment. | Optional | Access Request, Accounting Request START, Accounting Request STOP, Accounting Request interim update |
| 17 | 3GPP- IPv6-DNS-Servers | List of IPv6 addresses of DNS servers for an APN | Optional | Access Accept |

# Acronyms

| | |
|---|---|
| **AAA** | authentication, authorization, and accounting |
| **AAL5** | ATM Adaptation Layer |
| **ADM** | Add/Drop Multiplexer |
| **AH** | Authentication Header |
| **AMPS** | Advanced Mobile Phone System |
| **ANSI** | American National Standards Institute |
| **AP** | access point |
| **APN** | access point name |
| **ARIB** | Association of Radio Industries and Businesses |
| **ARP** | Address Resolution Protocol |
| **ARQ** | Automatic Repeat Request |

| | |
|---|---|
| **AS** | autonomous system |
| **ASP** | application service provider |
| **ATM** | Asynchronous Transfer Mode |
| **AuC** | Authentication Center |
| **AVP** | attribute-value pair |
| **BCP** | Best Current Practice |
| **BER** | bit error rate |
| **BG** | border gateway |
| **BGP** | Border Gateway Protocol |
| **BHCA** | busy hour call attempts |
| **BSC** | Base Station Controller |
| **BSS** | Base Station Subsystem |
| **BTS** | Base Transceiver Station |
| **CAMEL** | Customized Applications for Mobile Network Enhanced Logic |
| **CAP** | CAMEL Application Part |
| **CDMA** | Code-Division Multiple Access |
| **CdPA** | called party address |
| **CDR** | Charging Data Record |
| **CE** | customer edge |
| **CEPT** | Conference of European Posts and Telecommunications Administrations |

| | |
|---|---|
| **CF** | CompactFlash |
| **CGF** | Charging Gateway Function |
| **CgPA** | calling party address |
| **CHAP** | Challenge Handshake Authentication Protocol |
| **CK** | cipher key |
| **CLP** | Cell Loss Priority |
| **CN** | core network (UMTS) |
| **COPS** | Common Open Policy Service |
| **COPS-PR** | Common Open Policy Service for Provisioning |
| **CoS** | class of service |
| **CP** | control plane |
| **CPCS** | Common Part Convergence Sublayer |
| **CPE** | customer premise equipment |
| **CRC** | Cyclical Redundancy Check |
| **CRL** | certificate revocation list |
| **CS** | Circuit Switched |
| **CSCF** | Call State Control Function |
| **CSD** | circuit-switched data |
| **CSE** | CAMEL Service Environment |
| **DES** | Data Encryption Standard |
| **DHCP** | Dynamic Host Configuration Protocol |

| | |
|---|---|
| **DiffServ** | Differentiated Services |
| **DNS** | Domain Name System |
| **DoS** | denial of service |
| **DP** | detection point |
| **DPC** | Destination Point Code |
| **DRX** | discontinuous reception |
| **DS0** | Digital Signal Level 0 |
| **DSSS** | Direct-Sequence Spread Spectrum |
| **DWDM** | dense wavelength-division multiplexing |
| **EAP** | Extensible Authentication Protocol |
| **EDGE** | Enhanced Data Rates for GPRS Evolution |
| **EDP** | event detection point |
| **EGPRS** | Enhanced GPRS |
| **EIR** | Equipment Identity Register |
| **EMS** | Element Management System |
| **ESP** | Encapsulating Security Payload |
| **ETSI** | European Telecommunications Standards Institute |
| **ETSI-BRAN** | ETSI Broadband Access Network |
| **FA** | Foreign Agent |
| **FDD** | Frequency Division Duplex |
| **FDM** | frequency-division multiplexing |
| **FDMA** | Frequency Division Multiple Access |

| | |
|---|---|
| **FEC** | forwarding equivalence class |
| **FHSS** | Frequency Hopping Spread Spectrum |
| **FQDN** | fully qualified domain name |
| **FSK** | frequency-shift keying |
| **FSM** | finite state machine |
| **FTP** | File Transfer Protocol |
| **Ga** | interface between SGSN and CGF or GGSN and CGF |
| **Gb** | interface between the BSS and SGSN (2G GPRS) |
| **GBN** | GPRS backbone network |
| **GBS** | GPRS backbone system |
| **Gc** | interface between GGSN and HLR |
| **G-CDR** | GGSN-CDR |
| **GERAN** | GSM/EDGE Radio Access Network |
| **Gf** | interface between SGSN and EIR |
| **GGSN** | Gateway GPRS Support Node |
| **Gi** | reference point between the GGSN and external networks |
| **GMM** | GPRS Mobility Management |
| **Gn** | interface between GGSN and SGSN or between SGSNs within a PLMN |
| **Gp** | interface between GGSN and SGSN in different PLMNs |

| | |
|---|---|
| **G-PDU** | GTP Protocol Data Unit |
| **GPRS** | General Packet Radio Service |
| **GPRS-CSI** | GPRS CAMEL Subscription Information |
| **GPRS SCF** | GPRS Service Control Function |
| **GPRS SSF** | GPRS Service Switching Function |
| **Gr** | interface between SGSN and HLR |
| **GRE** | Generic Routing Encapsulation |
| **GRX** | GPRS Roaming Exchange |
| **Gs** | interface between SGSN and MSC/VLR |
| **GSM** | Global System for Mobile Communications |
| **GSM CCF** | GSM Service Control Function |
| **GSN** | GPRS Support Node |
| **GT** | global title |
| **GTP** | GPRS Tunneling Protocol |
| **GTP-C** | GPRS Tunneling Protocol-Control plane |
| **GTP-U** | GPRS Tunneling Protocol-User plane |
| **GTT** | Global Title Translation |
| **GUI** | graphical user interface |
| **HA** | Home Agent |
| **HLR** | Home Location RegisterHPLMN—Home PLMN |
| **HPMN** | Home Public Mobile Network |
| **HSCSD** | High-Speed Circuit-Switched Data |

| | |
|---|---|
| **HSS** | Home Subscriber Server |
| **HTTP** | Hypertext Transfer Protocol |
| **IBGP** | Internet Border Gateway Protocol |
| **ICMP** | Internet Control Message Protocol |
| **IE** | Information Element |
| **IEEE** | Institute of Electrical and Electronics Engineers |
| **IETF** | Internet Engineering Task Force |
| **IKE** | Internet Key Exchange |
| **IMEI** | International Mobile Equipment Identifier |
| **IMS** | IP Multimedia Subsystem |
| **IMSI** | International Mobile Station Identifier |
| **IP** | Internet Protocol |
| **IPCP** | Internet Protocol Control Protocol |
| **IPLMN** | Interrogating PLMN |
| **IPSec** | IP Security |
| **IPv4** | Internet Protocol Version 4.0 |
| **IPv6** | Internet Protocol Version 6.0 |
| **IREG** | International Roaming Expert Group |
| **ISD** | Insert Subscriber Data |
| **ISDN** | Integrated Services Digital Network |
| **ISP** | Internet service provider |
| **ITU** | International Telecommunications Union |

| | |
|---|---|
| **ITU-T** | ITU-Telecommunication Standardization Sector |
| **Iu-cs** | interface between the RNC and UMTS Circuit Core Network |
| **Iu-ps** | interface between the RNC and UMTS Packet Core Network |
| **Iur** | interface between RNCs |
| **IWF** | Interworking Function |
| **IW-MSC** | Interworking MSC |
| **IXC** | Internet Exchange |
| **L2F** | Layer Two Forwarding |
| **L2TP** | Layer Two Tunneling Protocol |
| **LA** | location area |
| **LAC** | L2TP Access Concentrator |
| **LAI** | Location Area Identifier |
| **LAN** | local area network |
| **LAU** | location area update |
| **LDAP** | Lightweight Directory Access Protocol |
| **LER** | label edge router |
| **LLC** | Logical Link Control |
| **LMDS** | Local Multipoint Distribution Services |
| **LNS** | L2TP Network Server |
| **LSP** | label switched path |
| **LSR** | label switching router |

| | |
|---|---|
| **MAC** | Media Access Control |
| **MAP** | GSM Mobile Application Part |
| **MCC** | Mobile Country Code |
| **MIB** | Management Information Base |
| **MM** | mobility management |
| **MN** | mobile node |
| **MNC** | Mobile Network Code |
| **MNRF** | Mobile Not Reachable Flag |
| **MNRG** | Mobile Not Reachable for GPRS Flag |
| **MNRR** | Mobile Not Reachable Reason |
| **MoU** | Memorandum of Understanding |
| **MPLS** | Multi-Protocol Label Switching |
| **MS** | mobile station |
| **MSC** | Mobile Switching Center |
| **MSISDN** | Mobile Station ISDN |
| **MT** | Mobile Terminal |
| **MTBF** | mean time between failures |
| **MTTR** | mean time to repair |
| **MVPN** | Mobile Virtual Private Network |
| **NAI** | Network Access Identifier |
| **NAPT** | Network Address Port Translation |
| **NAS** | network access server |

| | |
|---|---|
| **NAT** | Network Address Translation |
| **NAT-PT** | Network Address Translation-Protocol Translation |
| **NAT-T** | NAT-Traversal |
| **NCP** | Network Control Protocol |
| **NEBS** | National Equipment Building Specification |
| **NI** | Network Identifier |
| **NIC** | Network Interface Card |
| **NM** | network management |
| **NMT** | Nordic Mobile Telephone system |
| **NNI** | network-to-network interface |
| **N-PDU** | Network Layer Protocol Data Unit |
| **NSAPI** | Network Services Access Point Identifier |
| **NTP** | Network Time Protocol |
| **OA&M** | operations, administration, and maintenance |
| **OFDM** | orthogonal frequency-division multiplexing |
| **OHG** | Operators Harmonization Group |
| **OSA** | Open System Architecture |
| **OSI** | Open Systems Interconnection |
| **PANA** | Protocol for Carrying Authentication for Network Access |
| **PAP** | Password Authentication Protocol |
| **PC** | personal computer |

| | |
|---|---|
| **PCF** | Packet Control Function |
| **PCMCIA** | Personal Computer Memory Card International Association |
| **PCO IE** | Protocol Configuration Options Information Element |
| **P-CSCF** | Proxy CSCF |
| **PDCP** | Packed Data Convergence Protocol |
| **PDN** | packet data network |
| **PDP** | Packet Data Protocol |
| **PDSN** | Packet Data Serving Node |
| **PDU** | Protocol Data Unit |
| **PE** | provider edge |
| **PKI** | Public Key Infrastructure |
| **PLMN** | Public Land Mobile Network |
| **PMM** | Packet Mobility Management |
| **PN N-PDU** | Number Present Flag (in the GTP R99 header) |
| **PPF** | Paging Proceed Flag |
| **PPP** | Point-to-Point Protocol |
| **PPTP** | Point-to-Point Tunneling Protocol |
| **PRD** | Public Reference Document |
| **PS** | packet switched |
| **PSCN** | packet-switched core network |
| **PS-CP** | packet-switched control plane |

| | |
|---|---|
| **PSPDN** | packet-switched public data network |
| **PSTN** | Public Switched Telephone Network |
| **PS-UP** | packet-switched user plane |
| **P-TMSI** | Packet TMSI |
| **PVC** | Permanent Virtual Circuit |
| **QNC** | Quick Net Connect |
| **QoS** | quality of service |
| **R4** | UMTS Release 4 |
| **R5** | UMTS Release 5 |
| **R99** | UMTS Release 99 |
| **RA** | routing area |
| **RAB** | Radio Access Bearer |
| **RAC** | routing area code |
| **RADIUS** | Remote Authentication Dial-in User Service |
| **RAI** | Routing Area Identifier |
| **RAN** | radio access network |
| **RANAP** | Radio Access Network Application Part |
| **RAS** | remote access server |
| **RAU** | Routing Area Update |
| **RFC** | Request for Comments |
| **RIL3** | Radio Interface Layer 3 protocol |
| **RLC** | Radio Link Control |

| | |
|---|---|
| **RNC** | Radio Network Controller |
| **RPDU** | Relay Protocol Data Unit |
| **RR** | Radio Resource |
| **RRC** | Radio Resource Control |
| **RRP** | registration response |
| **RRQ** | registration request |
| **RTOS** | real-time operating system |
| **RTP** | Real-Time Protocol |
| **SA** | security association |
| **SAAL** | Signaling ATM Adaptation Layer |
| **SAD** | Security Association Database |
| **SAP** | service access point |
| **SCCP** | Signaling Connection Control Part |
| **S-CDR** | SGSN-CDR |
| **SCP** | service control point |
| **S-CSCF** | Serving CSCF |
| **SCTP** | Simple Control Transmission Protocol |
| **SDH** | Synchronous digital hierarchy |
| **SDO** | Standard Definition Organization |
| **SDU** | Service Data Unit |
| **SEP** | Signaling End Point |
| **SGSN** | Serving GPRS Support Node |

| | |
|---|---|
| **SIM** | Subscriber Identity Module |
| **SIP** | Session Initiation Protocol |
| **SLA** | service level agreement |
| **SLIP** | Serial Line Internet Protocol |
| **SMC** | Short Message Control (entity) |
| **SMC-GP** | Short Message Control-GPRS |
| **SM-CP** | Short Message Control Protocol |
| **SMG** | Special Mobile Group |
| **SMR** | Short Message Relay (entity) |
| **SM-RP** | Short Message Relay Protocol |
| **SMS** | Short Message Service |
| **SMS-CSI** | SMS CAMEL Subscription Information |
| **SMS-SC** | SMS Service Center |
| **SNDCP** | Sub-Network Dependent Convergence Protocol |
| **SNMP** | Simple Network Management Protocol |
| **SONET** | Synchronous Optical Network |
| **SPD** | Security Policy Database |
| **SRNC** | Serving Radio Network Controller |
| **SRNS** | Serving Radio Network Subsystem |
| **SSCF-NNI** | Service-Specific Coordination Function at the Network Node Interface |
| **SSCOP** | Service-Specific Connection-Oriented Protocol |

| | |
|---|---|
| **SSL** | Secure Socket Layer |
| **SSN** | Sub-System Number |
| **STP** | signal transfer point |
| **TACS** | Total Access Communication System |
| **TCP** | Transmission Control Protocol |
| **TDD** | Time Division Duplex |
| **TDM** | Time Division Multiplexing |
| **TDMA** | Time Division Multiple Access |
| **TE** | terminal equipment |
| **TEID** | Tunnel Endpoint Identifier |
| **TFT** | Traffic Flow Template |
| **TI** | Transaction Identifier |
| **TIA/EIA** | Telecommunications Industries Association/Electronic Industries Association |
| **TID** | Tunnel Identifier |
| **TLD** | Top Level Domain |
| **TLS** | Transport Level Security |
| **TMSI** | Temporary Mobile Subscriber Identifier |
| **T-PDU** | Payload of a GTP PDU |
| **TPDU** | Transfer Protocol Data Unit |
| **TRNC** | Target Radio Network Controller |
| **TSG** | Technical Standardization Group |

| | |
|---|---|
| **TTA** | Telecommunications Technology Association |
| **TTC** | Telecommunication Technology Committee |
| **UDP** | User Datagram Protocol |
| **UDR** | Usage Data Record |
| **UE** | user equipment |
| **UMTS** | Universal Mobile Telecommunication System |
| **UMTS-CS** | UMTS-Circuit-Switched Domain |
| **UMTS-PS** | UMTS-Packet-Switched Domain |
| **UNI** | user-to-network interface |
| **UP** | user plane |
| **URA** | UTRAN Registration Area |
| **USIM** | UMTS Subscriber Identification Module |
| **USSD** | Unstructured Supplementary Services Data |
| **UTRAN** | UMTS Terrestrial Radio Access Network |
| **Uu** | interface between MT and UTRAN |
| **UWCC** | Universal Wireless Communications Consortium |
| **V&V** | verification and validation |
| **VC** | virtual circuit |
| **VHE** | Virtual Home Environment |
| **VLR** | Visitor Location Register |
| **VoIP** | Voice over IP |

| | |
|---|---|
| **VPLMN** | Visited PLMN |
| **VPN** | Virtual Private Network |
| **VSA** | vendor-specific attribute |
| **WAN** | wide area network |
| **WAP** | Wireless Application Protocol |
| **WECA** | Wireless Ethernet Compatibility Alliance |
| **WEP** | Wireline Equivalent Protocol |
| **WG** | Working Group |
| **WISP** | Wireless Internet service provider |
| **WLAN** | Wireless LAN |
| **WWW** | World Wide Web |

# Bibliography

## Books

[Black2000] Black, U. 2000 *PPP and L2TP: Remote Access Communications.* Upper Saddle River, NJ: Prentice Hall.

[Comer1995] Comer, D. 1995. *Internetworking with TCP/IP, Volume 1.* Upper Saddle River, NJ: Prentice Hall.

[Doraswamy1999] Doraswamy, N., and D. Harkins. 1999. *IPSec: The New Security Standard for the Internet, Intranets, and Virtual Private Networks.* Upper Saddle River, NJ: Prentice Hall.

[Eberspacher2001] Eberspacher, J., H. Vogel, and C. Bettstetter. 2001. *GSM Switching, Services and Protocols.* West Sussex, England: Wiley & Sons.

[Garg2002] Garg, V. 2002. *Wireless Network Evolution: 2G to 3G.* Upper Saddle River, NJ: Prentice Hall.

[Hennessy1996] Hennessy, J., and P. Peterson. 1996. *Computer Architecture: A Quantitative Approach.* San Francisco: Morgan Kaufmann Publishers.

[Kaaranen2001] Kaaranen, H., A. Ahtiainen, L. Laitinen, S. Naghian, and V. Niemi. 2001. *UMTS Networks: Architecture, Mobility and Services.* West Sussex, England: Wiley & Sons.

[Kotler1999] Kotler, P. 2000. *Marketing Management.* Upper Saddle River, NJ: Prentice Hall.

[Solomon1998] Solomon, J. 1998. *Mobile IP.* Upper Saddle River, NJ: Prentice Hall.

[Yuan2001] Yuan, R., and T. Strayer. 2001. *Virtual Private Networks: Technologies and Solutions.* Upper Saddle River, NJ: Addison-Wesley.

## Papers and Articles

Gupta, V., and S. Gupta. 2001. "Securing the Wireless Internet." *IEEE Communications Magazine,* Volume 39, Issue 12. (December): pp. 68-74.

Juha Ala-Laurila, J., J. Mikkonen, and J. Rinnemaa. 2001. "Wireless LAN Access Network Architecture for Mobile Operators." *IEEE Communications Magazine,* Volume 39, Issue 11. (November): pp. 82-89.

McCann, P. and T. Hiller. 2001 "An Internet Infrastructure for Cellular CDMA Networks Using Mobile IP." *IEEE Personal Communications,* Volume 7, Issue 4. (August): pp. 26-32.

Park, J. 2002. "Wireless Internet Access for Mobile Subscribers Based on the GPRS/UMTS Network." *IEEE Communications Magazine,* Volume 40, Issue 4. (April): pp. 38-49.

Perkins, C. "Mobile IP joins forces with AAA." *IEEE Personal Communications,* Volume 7, Number 4. (August 2000): pp. 59-61.

Xu, S. and T. Saadawi. 2001. "Does 802.11 MAC Protocol Work Well in Multihop Wireless Ad Hoc Networks?" *IEEE Communications Magazine,* Volume 39, Issue 6. (June): pp. 130-137.

Zahariadis, T., K. Vaxevanakis, C. Tsantilas, N. Zervos, and N. Nikolaou. 2002. "Global Roaming in Next-Generation Networks." *IEEE Communications Magazine,* Volume 40, Issue 2. (February): pp. 145-151.

## Standards

[ETSI BRAN TR101.957] Broadband Radio Access Networks (BRAN); HIPERLAN Type 2; Requirements and Architectures for Interworking between HIPERLAN/2 and 3rd Generation Cellular systems.

[ITU-T Q.2210] ITU-T Recommendation Q.2110 "B-ISDN ATM Adaptation Layer-Service Specific Connection Oriented Protocol (SSCOP)," (7/1994).

[ITU-T Q.2100] ITU-T Recommendation Q.2100 "B-ISDN Signaling ATM Adaptation Layer (SAAL) - overview description," (7/1994).

[IS835] TIA/EIA/TR-45, IS-835. 2000. "Wireless IP Network Standard," June 2000.

[TSB115] TIA/EIA/TSB-115. 2001. "CDMA2000 Wireless IP Architecture Based on IETF Protocols," December 2000.

[IS95] TIA/EIA IS-95, "Mobile Station-Base Station Compatibility Standard for Dual-Mode Wideband Spread Spectrum Cellular System," 1996.

[TS 02.02] GSM 02.02, "Bearer services (BS) supported by GSM PLMN."

[TS 03.60] GSM 03.60, "GPRS Service Description, Stage 2."

[TS 02.34] GSM 02.34, "High-Speed Circuit Switched Data (HSCSD), Stage 1".

[IS2000] TIA/EIA/IS-2000.2-A, "Physical Layer Standard for CDMA2000 Standards for Spread Spectrum Systems," March 2000.

[IS2001] TIA/EIA/IS-2001, "Inter-operability Specification (IOSv4.0) for CDMA2000 Access Network Interfaces," 2001 (TIA TR45.4, PN-4545).

[IS707] TIA/EIA/IS-707-A2, "Data Service Options for Spread Spectrum Systems," July 1998.

[3GPP TS 24.008] 3GPP Technical Specification 24.008, "Mobile Radio Interface Layer 3 Specification; Core Network Protocols; Stage 3," 2002.

[3GPP TS 23.078] 3GPP Technical Specification 23.078, "Applications for Mobile Network Enhanced Logic (CAMEL) Phase 3 - Stage 2," 2001.

[3GPP TS 32.015] 3GPP Technical Specification 32.015, "Telecommunication Management; Charging and Billing; 3G Call and Event Data for the Packet Switched (PS) Domain," Release 99, 2001.

[3GPP TS 32.215] 3GPP Technical Specification 32.215, "Telecommunication Management; Charging and Billing; 3G call and event data for the Packet Switched (PS) domain," Release 4 and Release 5, 2002.

[3GPP TS 29.061] 3GPP Technical Specification 29.061, "Packet Domain; Interworking between the Public Land Mobile Network (PLMN). Supporting Packet Based Services and Packet Data Networks (PDN)," 2002.

[3GPP TS 27.060] 3GPP Technical Specification 27.060, "Packet Domain; Mobile Station (MS) Supporting Packet Switched Services," 2001.

[3GPP TS 29.060] 3GPP Technical Specification 29.060, "General Packet Radio Service (GPRS); GPRS Tunnelling Protocol (GTP) across the Gn and Gp Interface," 2001.

[TS 23.060] 3GPP Technical Specification 23.060, "General Packet Radio Service (GPRS); Service Description; Stage 2," 2002.

[3GPP TS 29.078] 3GPP Technical Specification 29.078, "Customized Applications for Mobile Network Enhanced Logic (CAMEL); CAMEL Application Part (CAP) Specification," 2002.

[3GPP TS 23.003] 3GPP Technical Specification 23.003, "Numbering, addressing and identification."

[PRD IR34] GSM Association International Roaming Expert Group, Public Reference Document IR.34 "Inter-PLMN Backbone Guidelines, " 2001.

[RFC1321]  "The MD5 Message-Digest Algorithm," 1992 RFC 1321.

[RFC1334]  "PPP Authentication Protocols," 1992 RFC 1334.

[RFC1661]  "The Point-to-Point Protocol (PPP)," 1999 RFC 1661.

[RFC1701]  "Generic Routing Encapsulation (GRE)," 1994 RFC 1701.

[RFC1918]  "Address Allocation for Private Internets," 1996 RFC 1918.

[RFC1974]  "PPP Stac LZS Compression Protocol," 1996 RFC 1974.

[RFC1994]  "PPP Challenge Handshake Authentication Protocol (CHAP)," 1996 RFC 1994.

[RFC2002]  "IP Mobility Support," 1996 RFC 2002.

[RFC2003]  "IP Encapsulation within IP," 1996 RFC 2003.

[RFC2004]  "Minimal Encapsulation within IP," 1996 RFC 2004.

[RFC2118]  "Microsoft Point-to-Point Compression (MPPC) Protocol," 1997 RFC 2118.

[RFC2131]  "Dynamic Host Configuration Protocol," 1997 RFC 2131.

[RFC2138]  "Remote Authentication Dial In User Service (RADIUS)," 1997 RFC 2138.

[RFC2139]  "RADIUS Accounting," 1997 RFC 2139.

[RFC2246]  "The TLS Protocol Version 1.0," 1999 RFC 2246.

[RFC2284]  "PPP Extensible Authentication Protocol (EAP)," 1998 RFC 2284.

[RFC2394]  "IP Payload Compression Using DEFLATE," 1998 RFC 2394.

[RFC2398]  "Some Testing Tools for TCP Implementors," 1998 RFC 2398.

[RFC2401]  "Security Architecture for the Internet Protocol," 1998 RFC 2401.

[RFC2402]  "IP Authentication Header," 1998 RFC 2402.

[RFC2406]  "IP Encapsulating Security Payload (ESP)," 1998 RFC 2406.

[RFC2409]  "The Internet Key Exchange (IKE)," 1998 RFC 2409.

[RFC2474]  "Definition of the Differentiated Services Field (DS 1998 Field) the IPv4 and IPv6 Headers," 1998 RFC 2474.

[RFC2475]  "An Architecture for Differentiated Services," 1998 RFC 2475.

[RFC2436]  "Collaboration between ISOC/IETF and ITU-T," 1998 RFC 2436.

[RFC2486]  "The Network Access Identifier," 1999 RFC 2486.

[RFC2547]  "BGP/MPLS VPNs," 1999 RFC 2547.

[RFC2597]  "Assured Forwarding PHB Group," 1999 RFC 2597.

[RFC2697]  "Single Rate Three Color Marker," 1999 RFC 2697.

[RFC2698]  "A Two Rate Three Color Marker," 1999 RFC 2698.

[RFC2709]  "Security Model with Tunnel-mode IPSec for NAT 1999 Domains," RFC 2709.

[RFC2740]  "OSPF for IPv6,"1999 RFC 2740.

[RFC2784]  "Generic Routing Encapsulation (GRE)," 2000 RFC 2784.

[RFC2820]  "Access Control Requirements for LDAP," 2000 RFC 2820.

[RFC2829]  "Authentication Methods for LDAP," 2000 RFC 2829.

[RFC2865] "Remote Authentication Dial In User Service (RADIUS)," 2000 RFC 2865.

[RFC2866] "RADIUS Accounting," 2000 RFC 2866.

[RFC2890] "Key and Sequence Number Extensions to GRE," 2000 RFC 2890.

[RFC2960] "Stream Control Transmission Protocol," 2000 RFC 2960.

[RFC2983] "Differentiated Services and Tunnels," 2000 RFC 2983

[RFC3022] "Traditional IP Network Address Translator 2001 (Traditional NAT)," 2001 RFC 3022.

[RFC3032] "MPLS Label Stack Encoding," 2001 RFC 3032.

[RFC3034] "Use of Label Switching on Frame Relay Networks Specification," 2001 RFC 3034.

[RFC3035] "MPLS using LDP and ATM VC Switching," 2001 RFC 3035.

[RFC3036] "LDP Specification," 2001 RFC 3036.

[RFC3084] "COPS Usage for Policy Provisioning (COPS-PR)," 2001 RFC 3084.

[RFC3118] "Authentication for DHCP Messages," 2001 RFC 3118.

[RFC3141] "CDMA2000 Wireless Data Requirements for AAA," 2001 RFC 3141.

[RFC3212] "Constraint-Based LSP Setup using LDP," 2002 RFC 3212.

[RFC3220] "IP Mobility Support for IPv4," 2002 RFC 3220.

# Index